Statistical Analysis of Cost-effectiveness Data

STATISTICS IN PRACTICE

Advisory Editors

Stephen Senn
University of Glasgow, UK

Marian Scott
University of Glasgow, UK

Peter Bloomfield
North Carolina State University, USA

Founding Editor

Vic Barnett
Nottingham Trent University, UK

Statistics in Practice is an important international series of texts which provide detailed coverage of statistical concepts, methods and worked case studies in specific fields of investigation and study.

With sound motivation and many worked practical examples, the books show in down-to-earth terms how to select and use an appropriate range of statistical techniques in a particular practical field within each title's special topic area.

The books provide statistical support for professionals and research workers across a range of employment fields and research environments. Subject areas covered include medicine and pharmaceutics; industry, finance and commerce; public services; the earth and environmental sciences, and so on.

The books also provide support to students studying statistical courses applied to the above areas. The demand for graduates to be equipped for the work environment has led to such courses becoming increasingly prevalent at universities and colleges.

It is our aim to present judiciously chosen and well-written workbooks to meet everyday practical needs. The feedback of views from readers will be most valuable to monitor the success of this aim.

A complete list of titles in this series appears at the end of the volume.

Statistical Analysis of Cost-effectiveness Data

Andrew R. Willan
University of Toronto, Canada

Andrew H. Briggs
University of Glasgow, UK

John Wiley & Sons, Ltd

Other Wiley Editorial Offices

John Wiley & Sons Inc., 111 River Street, Hoboken, NJ 07030, USA

Jossey-Bass, 989 Market Street, San Francisco, CA 94103-1741, USA

Wiley-VCH Verlag GmbH, Boschstr. 12, D-69469 Weinheim, Germany

John Wiley & Sons Australia Ltd, 42 McDougall Street, Milton, Queensland 4064, Australia

John Wiley & Sons (Asia) Pte Ltd, 2 Clementi Loop #02-01, Jin Xing Distripark,
Singapore 129809

John Wiley & Sons Canada Ltd, 22 Worcester Road, Etobicoke, Ontario, Canada M9W 1L1

Wiley also publishes its books in a variety of electronic formats. Some content that appears
in print may not be available in electronic books.

Library of Congress Cataloging-in-Publication Data

(to follow)

British Library Cataloguing in Publication Data

A catalogue record for this book is available from the British Library

ISBN-13 978-0-470-85626-0 (HB)
ISBN-10 0-470-85626-2 (HB)

Typeset in 11/13pt Photina by TechBooks, New Delhi, India

For Bernie

*All author proceeds donated
to the Emma and Lucy O'Brien Education Fund.*

Contents

Preface   xi

1   Concepts   1

   1.1   Introduction   1
   1.2   Cost-effectiveness data and the parameters of interest   2
   1.3   The cost-effectiveness plane, the ICER and INB   5
   1.4   Outline   8

2   Parameter Estimation for Non-censored Data   11

   2.1   Introduction   11
   2.2   Cost   12
      2.2.1   Sample means for estimating incremental cost   12
      2.2.2   Using multiple regression models   14
      2.2.3   Transformation (and the retransformation problem)   15
      2.2.4   Generalized linear models   17
      2.2.5   Two-part models for excess zeros   18
      2.2.6   Cost prediction models   19
   2.3   Effectiveness   20
      2.3.1   Probability of surviving   21
      2.3.2   Mean survival time   21
      2.3.3   Mean quality-adjusted survival time   22
      2.3.4   Mean quality-adjusted survival: controlling for baseline utility   24
   2.4   Summary   25

3   Parameter Estimation for Censored Data   27

   3.1   Introduction   27
   3.2   Mean Cost   28
      3.2.1   Direct (Lin) method   29
      3.2.2   Inverse-probability weighting   31
   3.3   Effectiveness   34
      3.3.1   Probability of surviving   34

3.3.2   Mean survival time   36
3.3.3   Mean quality-adjusted survival time   39
3.4   Summary   42

4   Cost-effectiveness Analysis   43

4.1   Introduction   43
4.2   Incremental cost-effectiveness ratio   44
4.3   Incremental net benefit   49
4.4   The cost-effectiveness acceptability curve   51
4.5   Using bootstrap methods   54
4.6   A Bayesian incremental net benefit approach   57
4.7   Kinked thresholds   60
4.8   Summary   64

5   Cost-effectiveness Analysis: Examples   67

5.1   Introduction   67
5.2   The CADET-Hp trial   67
5.3   Symptomatic hormone-resistant prostate cancer   72
5.4   The Canadian implantable defibrillator study (CIDS)   77
5.5   The EVALUATE trial   82
5.6   Bayesian approach applied to the UK PDS study   86
5.7   Summary   90

6   Power and Sample Size Determination   93

6.1   Introduction   93
6.2   Approaches based on the cost-effectiveness plane   94
       6.2.1   Briggs and Gray   95
       6.2.2   Willan and O'Brien   98
       6.2.3   Gardiner *et al.*   101
6.3   The classical approach based on net benefit   103
       6.3.1   The method   103
       6.3.2   Example   105
6.4   Bayesian take on the classical approach   106
       6.4.1   The Method   106
       6.4.2   Example   107
6.5   The value of information approach   108
       6.5.1   The method   108
       6.5.2   Example   114
6.6   Summary   116

7   Covariate Adjustment and Sub-group Analysis   117

7.1   Introduction   117
7.2   Non-censored data   118
       7.2.1   Example, non-censored data   121

7.3 Censored data 129
    7.3.1 Cost 131
    7.3.2 Quality-adjusted survival time 132
    7.3.3 Survival time 134
    7.3.4 The Canadian implantable defibrillator study (CIDS) 135
    7.3.5 The evaluate trial 138
7.4 Summary 142

8 Multicenter and Multinational Trials 145

8.1 Introduction 145
8.2 Background to multinational cost-effectiveness 147
8.3 Fixed effect approaches 151
    8.3.1 Willke *et al.* 151
    8.3.2 Cook *et al.* 152
8.4 Random effects approaches 154
    8.4.1 Aggregate level analysis: multicenter trials 154
    8.4.2 Aggregate level analysis: multinational trials 156
    8.4.3 Hierarchical modeling 162
8.5 Summary 164

9 Modeling Cost-effectiveness 165

9.1 Introduction 165
9.2 A general framework for modeling cost-effectiveness results 166
9.3 Case study: an economic appraisal of the goal study 167
    9.3.1 The GOAL study 168
    9.3.2 Standard approach to estimating cost-effectiveness 170
    9.3.3 An alternative approach to estimating cost-effectiveness 171
    9.3.4 Comparing the two analyses of GOAL 179
9.4 Summary 180

References 183

Author Index 193

Subject Index 195

Series List 197

Preface

This book describes statistical methods applied to cost-effectiveness analysis. It represents the experience over many years of the author's involvement in the application and methodology of health economic evaluation. The focus on randomised clinical trials reflects the fact that the trend towards collecting not only clinical, but also economic, data alongside clinical trials was the driving force behind many of the methodological developments described in the text. Health economics is a relatively young discipline and the use of clinical trials as a vehicle for economic evaluations began in earnest only twenty years ago. As a consequence, there has been a high degree of methodological development since then, with most of the reporting confined to journal articles. The aim of this book is to draw together those developments in a single source which we hope will be of interest to students of statistics, keen to understand more about health economics, and students of health economics, keen to understand the statistical methods required for undertaking economic evaluation of health care interventions. The exposition is at a technical level roughly equivalent to that found in final year undergraduate mathematics and statistics courses or postgraduate social sciences courses.

The book itself naturally divides into two parts. The first part (up to Chapter 5) deals with the established approach for the presentation of cost-effectiveness analyses, with a focus on estimating health outcomes and resource use costs. The second part of the book (Chapters 6 through 9) handles specific issues in more depth to give a fuller understanding of the nuances of a modern cost-effectiveness analysis where patient-level data are available.

In the preparation of any book there are numerous colleagues and students who have provided the inspiration and insight, as well as friends and family who have provided the encouragement and support, necessary to bring such a project to fruition. We are extremely grateful to all those people who have helped us over the years and aided us to a greater or lesser extent in supporting our endeavours and correcting our mistakes. However, one person stands out as the true inspiration for this book. A friend and colleague who had a major influence on both of our careers in the area of health economic evaluation, albeit in different ways, Bernie O'Brien was a rare person – someone with a keen intellect, an infectious enthusiasm, and a generosity of ideas that could not fail to rub off on those around him. His untimely death on the 13th of February, 2004 was a terrible shock and leaves a vacuum in the health economics community, as well as for his wife Karen and daughters, Emma and Lucy. We dedicate this book to Bernie's memory.

1

Concepts

1.1 INTRODUCTION

There is a growing expectation from health care policymakers that evidence supporting the cost-effectiveness of new health care interventions, particularly pharmaceuticals, be provided along with the customary data on efficacy and safety. In Australia (Commonwealth of Australia, 1990) and Canada (Detsky, 1993) there are formal requirements that pharmaceutical companies present evidence of cost-effectiveness before a drug is granted reimbursement status on a formulary. In the United States there is demand for such economic data from third-party insurers, see Leaf (1989).

There are two general approaches to performing an economic evaluation of a health care intervention, see O'Brien (1996). One approach combines the efficacy and safety data from randomized clinical trials (RCTs) with cost data from secondary, non-trial sources in a decision analysis model. In such models the problem of inferential uncertainty is addressed using sensitivity analyses to determine what effect varying the model assumptions has on the results, see Briggs *et al.* (1994). The other approach uses health care utilization data collected on individual patients prospectively as part of an RCT. The health care utilization data combined with the appropriate price weights yield a measure of cost for each patient. Measuring effectiveness and cost at the patient level permits the use of more conventional methods of statistical inference to quantify the uncertainty due to sampling and measurement error. Since the early 1990s, when such data became

Statistical Analysis of Cost-effectiveness Data. A. Willan and A. Briggs
© 2006 John Wiley & Sons, Ltd.

more common, numerous articles have been published in the area of the statistical analysis of cost-effectiveness data. Initially, efforts were concentrated on providing confidence intervals for incremental cost-effectiveness ratios, but more recently, due to concerns regarding ratio statistics, the concept of incremental net benefit has been proposed as an alternative.

The purpose of this book is to provide an illustrated summary of some of the key developments published in the last 10 years that deal with statistical issues related to the cost-effectiveness comparison of two groups when measures of effectiveness and cost are observed at the subject level. The context used throughout the book is that of patients in a two-arm RCT where patients are randomized to Treatment (T) or Standard (S), but the methods apply to the comparison of any two groups, subject to the concerns one might have regarding bias due to the lack of random group allocation.

1.2   COST-EFFECTIVENESS DATA AND THE PARAMETERS OF INTEREST

In a cost-effectiveness analysis (CEA), whether an incremental cost-effectiveness ratio (ICER) or an incremental net benefit approach is taken, five parameters need to be estimated. Two of the parameters are the differences between treatment arms of mean effectiveness and costs, denoted by Δ_e and Δ_c, respectively. The other three parameters are the variances and covariance of those estimators. With the estimators of these five parameters, a CEA, based on either the incremental cost-effectiveness ratio or incremental net benefit, can be performed. For non-censored data the estimators are simple functions of the sample means, variances and covariance. For censored data estimation procedures are decidedly more complex.

Typically, the measure of effectiveness in a CEA is associated with a clinical event experienced by the patient. Quite commonly the event is death, but it could be relapse or reaching a pre-specified level of symptom relief. For simplicity, unless otherwise noted, we assume that the event is death. The simplest measure of effectiveness based on event data is the probability of the event not occurring within a specified period of time from randomization. The specified period of time is often referred to as the duration of interest and denoted by τ. Let the random variable D_{Ti} be the time from randomization to the event for the ith patient on arm T and let $S_T(t) = \Pr(D_{Ti} \geq t)$, then the measure of effectiveness is given by $S_T(\tau)$ and denoted as π_T. $S_T(t)$ is the survival function for patients on arm T. Defining D_{Si}, $S_S(t)$ and π_S similarly for patients on arm S, the parameter of interest for effectiveness, denoted Δ_e, is given by

$$\Delta_e = S_T(\tau) - S_S(\tau) = \pi_T - \pi_S \qquad (1.1)$$

The quantity Δ_e is the absolute risk reduction, and $1/\Delta_e$ is the mean number of patients that need to be treated with T rather than S to prevent a death. This quantity is usually referred to informally as the 'number-needed-to-treat' or more simply as the NNT. If the probability of surviving 5 years is 0.6 for patients on arm T and only 0.5 for patients on arm S, then we say that 10 (i.e. 1/0.1) patients need to be treated with T rather than S to prevent one death, or more simply that the NNT is 10.

Another measure of effectiveness based on event data is the mean survival time over the duration of interest, otherwise referred to as the restricted mean survival time. The restricted mean survival time is sensitive to the entire survival curve from 0 to τ, and not just its value at τ. The restricted mean survival time for a particular arm, denoted as μ_j, $j = T, S$, is the area under the respective survival curve from 0 to τ, i.e.

$$\mu_j = \int_0^\tau S_j(t)dt$$

and the parameter of interest to be estimated for effectiveness is given by

$$\Delta_e = \int_0^\tau S_T(t)dt - \int_0^\tau S_S(t)dt = \mu_T - \mu_S \qquad (1.2)$$

For mean survival time quantity $1/\Delta_e$ is the NNT to gain one year of life over the duration of interest. If the restricted mean survival time over 5 years for a patient on arm T is 4 and only 3.75 for a patient on arm S, then the NNT to gain one year of life is 4 (i.e. $1/0.25$). The mean total restricted survival time of 4 patients on arm T is $4 \times 4 = 16$, while on arm S the mean total restricted survival time of 4 patients is $4 \times 3.75 = 15$.

The third measure of effectiveness based on survival data is the mean quality-adjusted survival time over the duration of interest, otherwise referred to as the restricted mean quality-adjusted survival time. Quality-adjusted survival time is based on the concept that patients experience, at any given time, a certain quality of life based on a utility scale for which 1 is perfect health and 0 is death, see Torrance (1986). Negative values are used to allow for states of health considered worse than death. If the quality of life at time t for a patient on a particular treatment arm is given by $Q_j(t)$, $j = T$, S, then the restricted mean quality-adjusted survival time is given by

$$\varphi_j \equiv \int_0^\tau Q_j(t)\mathrm{d}t$$

and the parameter of interest to be estimated for effectiveness is given by

$$\Delta_e = \int_0^\tau Q_T(t)\mathrm{d}t - \int_0^\tau Q_S(t)\mathrm{d}t = \varphi_T - \varphi_S \qquad (1.3)$$

For quality-adjusted survival time the quantity $1/\Delta_e$ is the NNT to gain one quality-adjusted life-year over the duration of interest.

The quantity Δ_e is the difference between treatment arms with respect to effectiveness, and is a different function of the survival curves, depending on which measure of effectiveness is of interest.

If we let v_j be the mean cost over the duration of interest for a patient in arm j, $j = T$, S, then the parameter of interest to be estimated for cost is given by

$$\Delta_c = v_T - v_S \qquad (1.4)$$

The observed cost for a given patient is simply the sum of the amounts of each resource consumed by the patient multiplied by the respective price weight. Which resources are included depends on the perspective taken by the analysis. If the analysis takes the perspective of the health care system, only resources covered under the system would be included. However, if a broader societal perspective were taken, then costs not covered under the system and items such as time lost from work and care by a family member could also be included; for a fuller discussion the reader is referred to Drummond *et al.* (1997). Estimation methods for Δ_e, Δ_c and the corresponding variances and covariance are given in Chapter 2 for non-censored data and in Chapter 3 for censored data.

1.3   THE COST-EFFECTIVENESS PLANE, THE ICER AND INB

Researchers have long used the cost-effectiveness plane to explore the policy interpretation of cost-effectiveness analyses. The cost-effectiveness (CE) plane is a graph with Δ_c and Δ_e plotted on the vertical axis and horizontal axis, respectively, as illustrated in Figure 1.1. For more discussion on the cost-effectiveness plane the reader is referred to Black (1990). Let $\Delta = (\Delta_e, \Delta_c)^T$. If, for a particular Treatment/Standard comparison, the point Δ is located in the Southeast (SE) quadrant (i.e. $\Delta_e > 0$, $\Delta_c < 0$), Treatment is said to dominate Standard because it is more effective and less costly, and the argument to adopt it to replace Standard is self-evident. By contrast, if Δ lies in the Northwest (NW) quadrant (i.e. $\Delta_e < 0$, $\Delta_c > 0$) Treatment is dominated by Standard, and its rejection as a replacement for Standard is the rational policy choice. It is in the Northeast (NE) and Southwest (SW) quadrants, referred to as the trade-off quadrants, that the magnitudes of Δ_e and Δ_c need to be considered to determine if Treatment is cost-effective.

To assist in this determination researchers have traditionally used the incremental cost-effectiveness ratio. The ICER is defined as $R \equiv \Delta_c/\Delta_e$, but can be written as

$$\frac{1}{\Delta_e}\Delta_c = \text{NNT} \times \Delta_c$$

It is easy to see then that the ICER is the product of the number of patients that need to be given Treatment to achieve an extra unit

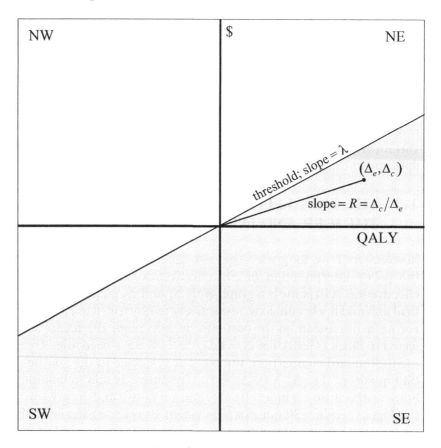

Figure 1.1 The cost-effectiveness plane

of effectiveness and the incremental cost of treating each of those patients, and is therefore the incremental cost of achieving a unit of effectiveness from using Treatment rather that Standard. On the CE plane the ICER is the slope of the line between the origin and the point Δ, see Figure 1.1. If the measure of effectiveness is the probability of surviving then the ICER is cost of saving a life (or preventing a death). If the measure of effectiveness is mean survival or mean quality-adjusted survival, then the ICER is the cost of achieving an extra year or quality-adjusted year of life (QALY), respectively. Essentially the ICER is the cost of an additional unit of effectiveness if Treatment is adopted over

Standard. This, as in any transaction, needs to be compared with what a policymaker is willing to pay.

The amount a policymaker is willing to pay is referred to as the willingness-to-pay (WTP), and is denoted by λ. The concept of WTP is discussed by Pauly (1995), and methods for quantifying it can be found in O'Brien and Gafni (1996), Johnson *et al.* (1998) and Hanley *et al.* (2003). By drawing a line through the origin with slope λ, the CE plane can be divided into two regions. For convenience this line will be referred to as the threshold. For points on the plane below and to the right of the threshold (the shaded area in Figure 1.1), Treatment is considered cost-effective, but for those above and to the left it is not. Since λ is positive, points in the SE quadrant are always below the threshold and therefore correspond to comparisons for which Treatment is cost-effective. On the other hand, points in the NW are always above the threshold and correspond to comparisons for which Treatment is not cost-effective. It is in the NE and SW quadrants that the concept of WTP allows for trade-off between effectiveness and costs. In the NE quadrant the slope of any point below the line is less than λ, i.e. $\Delta_c/\Delta_e < \lambda$ which implies that $\Delta_c < \Delta_e\lambda$. Therefore, the increase in value ($\Delta_e\lambda$) is greater than the increase in cost, making Treatment cost-effective. In the SW quadrant the slope of any point below the line is greater than λ, and since Δ_e and Δ_c are both negative (i.e. treatment is less effective and less costly), we have $\Delta_c/\Delta_e = |\Delta_c|/|\Delta_e| > \lambda$ which implies that $|\Delta_c| > |\Delta_e\lambda|$. Therefore, the value lost ($|\Delta_e\lambda|$) is less than the amount saved ($|\Delta_c|$), making Treatment cost-effective. In summary, Treatment is cost-effective if

$$\text{A}: \quad \frac{\Delta_c}{\Delta_e} < \lambda \quad \text{if } \Delta_e > 0; \quad \text{or} \quad \frac{\Delta_c}{\Delta_e} > \lambda \quad \text{if } \Delta_e < 0 \quad (1.5)$$

Expression (1.5) (Hypothesis A) defines the region below the threshold and can be thought of as the alternative hypothesis for the null Hypothesis H, given by:

$$\text{H}: \quad \frac{\Delta_c}{\Delta_e} \geq \lambda \quad \text{if } \Delta_e > 0; \quad \text{or} \quad \frac{\Delta_c}{\Delta_e} \leq \lambda \quad \text{if } \Delta_e < 0 \quad (1.6)$$

Rejecting H in favour of A would provide evidence to adopt Treatment. These expressions are somewhat awkward and can be simplified considerably by the introduction of incremental net benefit.

The incremental net benefit (INB) is a function of λ, and is defined as

$$b_\lambda \equiv \Delta_e \lambda - \Delta_c \qquad (1.7)$$

b_λ is the incremental net benefit because it is the difference between incremental value $(\Delta_e \lambda)$ and incremental cost (Δ_c). Treatment is cost-effective if, and only if, $b_\lambda > 0$, regardless of the sign of Δ_e. To see this, both inequalities involving the ICER in Expression (1.5) can be rearranged to the inequality $\Delta_e \lambda - \Delta_c > 0$. Similarly, both inequalities involving the ICER in Expression 1.6 can be rearranged to the inequality $\Delta_e \lambda - \Delta_c \leq 0$. Therefore, in terms of INB the null and alternative hypotheses become

$$\text{H} : \Delta_e \lambda - \Delta_c \leq 0 \qquad \text{versus} \qquad \text{A} : \Delta_e \lambda - \Delta_c > 0 \qquad (1.8)$$

On the CE plane b_λ is the vertical distance from the point Δ to the threshold, being positive if it is below the line and negative otherwise. Because it has slope λ, the point on the threshold with abscissa equal to Δ_e is $(\Delta_e, \Delta_e \lambda)$ and so the vertical distance between it and Δ is $\Delta_e \lambda - \Delta_c$, see Figure 1.2.

The incremental net health benefit (INHB) is defined as $\Delta_e - \Delta_c/\lambda = b_\lambda/\lambda$ and measures net benefit in units of effectiveness. Since INHB is simply a positive constant times INB, statistical inference made on one will be identical to statistical inference made on the other. INB has the advantage of being linear in λ. Therefore in a sensitivity analysis of WTP, the plots of INB by λ, are straight lines. Another advantage of INB is that it generalizes to more than one outcome. In a trial of patients at risk of thrombosis, if Δ_{e1}, Δ_{e2} and Δ_{e3} are the differences of the probability of avoiding death, thrombosis and stroke, respectively, and if λ_1, λ_2 and λ_3 are the corresponding WTP values, then INB is defined as $\Delta_{e1}\lambda_1 + \Delta_{e2}\lambda_2 + \Delta_{e3}\lambda_3 - \Delta_c$. A corresponding formulation in INHB is not possible.

1.4 OUTLINE

The remainder of the book is organized as follows. Methods for estimating Δ_e and Δ_c and their variances and covariances for non-censored data are given in Chapter 2. The methods make use of simple statistics,

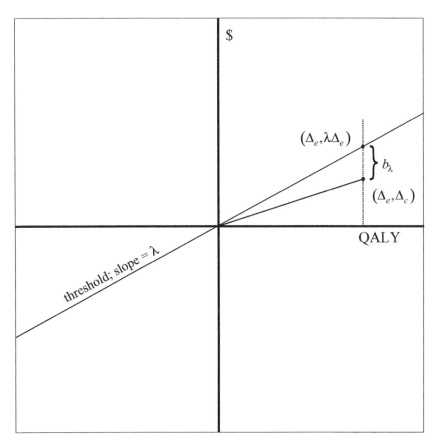

Figure 1.2 INB on the cost-effectiveness plane

such as proportions and sample means and variances. Estimation methods for censored data are given in Chapter 3. The methods include life-table procedures, the direct method of Lin *et al.* (1997) and inverse probability weighting. How the parameters are used in a cost-effectiveness analysis is described in Chapter 4. Emphasis is placed on estimating the ICER and INB, along with their confidence limits, and constructing cost-effectiveness acceptability curves. In Chapter 5 the methods of Chapters 2, 3 and 4 are illustrated with examples. Methods for determining sample sizes, using both classical and Bayesian approaches, are given in Chapter 6. In Chapter 7 regression methods for covariate adjustment and testing for treatment by prognostic factor interactions are described, along with several examples. The issues regarding multicenter and multinational trials are the subject of Chapter 8. In Chapter 9 a more general framework of statistical modeling is proposed, which is based on modeling the separate components cost-effectiveness to build indirect estimates of incremental cost and effectiveness.

2

Parameter Estimation for Non-censored Data

2.1 INTRODUCTION

Data from a clinical trial are said to be non-censored if all patients were followed either until death or for the total duration of interest. This implies that no patients were lost to follow-up and that the delay in time between when the last patient was randomized and when the analysis was conducted was sufficiently long that all patients were followed for the duration of interest. Consequently, the measure of effectiveness and total cost are known for all patients. Trials of treatments for acute health events, such as myocardial infarction and asthma exacerbations, are likely to have non-censored data since the duration of interest is usually relatively short. As the duration lengthens, however, the probability that some patients will be lost to follow-up increases. In addition, for trials with long durations of interest and staggered patient entry, analysis will likely occur prior to all patients having been followed for the entire duration of interest, resulting in administrative censoring.

The statistical methods for estimating the five parameters required to conduct a cost-effectiveness analysis when the data are non-censored are given in the remaining sections of this chapter. For brevity, in the remainder of the chapter and throughout the rest of the book the terms 'mean cost' or 'mean effectiveness' will be used in place of the term 'restricted (to the duration of interest) mean'.

Statistical Analysis of Cost-effectiveness Data. A. Willan and A. Briggs
© 2006 John Wiley & Sons, Ltd.

Estimation of the between-treatment difference in mean cost is dealt with in the Section 2.2, beginning with the common approach in clinical trials of using the difference in sample means. However, in practice cost data can be skewed and will often have excess zeros, which can lead to inefficiency in the standard least-squares approach. The most common approaches advocated for overcoming these problems are outlined in the remainder of the section. All three functions of survival—probability of surviving, mean survival time and mean quality-adjusted survival time—are considered in Section 2.3. Particular attention is given to the importance of controlling for baseline utility in the estimation of QALY gains. Because the data are non-censored, only simple statistics, such as proportions and sample means, variances and covariances, are required. The methods for handling censoring are the subject of Chapter 3.

2.2   COST

Incremental cost is often estimated by the between-treatment difference in the sample means. However, the nature of health service resource use is that cost data are often highly skewed and can exhibit a large proportion of zero values. Nevertheless, the fundamental interest of the economist in means was outlined in the Chapter 1 and leads to a rejection of rank order statistics, such as the median, and associated rank sum tests for location, when analyzing such data. Possible solutions include the use of transformation and smearing, the direct transformation of the expectation of cost and the use of two-part models for separating out zero observations.

2.2.1   Sample means for estimating incremental cost

For non-censored data the total cost over the duration of interest is observed for all patients, and the mean cost in each arm of the trial is often estimated by the sample mean. If we let C_{ji} be the observed total cost over the duration of interest for the jith patient, where the jith patient is the ith patient on the jth arm, where $j = T, S; i =$

$1, 2, \ldots n_j$; and n_j is the number of patients randomized to the jth arm, then v_j, the mean cost for the jth arm, is estimated by

$$\hat{v}_j = \frac{1}{n_j} \sum_{i=1}^{n_j} C_{ji}$$

Therefore, the between-arm difference in mean cost is estimated by

$$\hat{\Delta}_c = \hat{v}_T - \hat{v}_S = \frac{1}{n_T} \sum_{i=1}^{n_T} C_{Ti} - \frac{1}{n_S} \sum_{i=1}^{n_S} C_{Si} \tag{2.1}$$

The variance of $\hat{\Delta}_c$ is estimated by

$$\hat{V}\left(\hat{\Delta}_c\right) = \hat{V}\left(\hat{v}_T\right) + \hat{V}\left(\hat{v}_S\right)$$
$$= \frac{1}{n_T(n_T - 1)} \sum_{i=1}^{n_T} (C_{Ti} - \hat{v}_T)^2 + \frac{1}{n_S(n_S - 1)} \sum_{i=1}^{n_S} (C_{Si} - \hat{v}_S)^2$$

$$\tag{2.2}$$

The formulation in Equation (2.2) allows the between-patient variance in cost to differ between arms.

Because of the right-skewing which is usually present in cost data, criticism is often leveled at the use of least-squares methods such as sample means and variances, see O'Hagan and Stevens (2003), Briggs and Gray (1998a), Thompson and Barber (2000), Nixon and Thompson (2005) and Briggs *et al.* (2005), and transformations, such as the logarithm and square root, are sometimes proposed. However, such transformations provide estimates on a scale not relevant to decision makers, see Manning and Mullahy (2001) and Thompson and Barber (2000).

One approach to this issue is to consider 'rules of thumb' for when skewness may cause concern for standard least-squares methods of analysis. Although it is commonly considered that the *central limit theorem* applies for samples with greater than 30 observations, ensuring a normal distribution of the sample mean whatever the distribution in the underlying population, this rule of thumb may not apply to non-symmetric distributions typical of cost data. An alternative rule of thumb for situations where the 'principal deviation from normality consists of marked positive skewness' is provided by Cochran (1977) with the suggestion that $n > 25\eta^2$ where η is the skewness coefficient in the sample and n is the sample size. The guideline was devised such that a 95% confidence interval will have an error probability no greater than 6%.

Additionally, a number of investigations into the issue of skewed data, using mostly simulated data, have drawn the conclusion that least-squares methods provide valid estimates of mean cost and the between-arm difference in mean cost. Lumley *et al.* (2002) provide a review of such investigations. Willan and O'Brien (1996) and Willan, Briggs and Hoch (2004) address specifically the issue of skewed cost data and find that, even when cost data is distributed as log-normal to the base 10, the distribution of $\hat{\Delta}_c$ exhibits little skewness and kurtosis for sample sizes as small as 25 per arm.

Nonetheless, the blind application of sample means and variances to cost data with extreme outliers is likely to lead to misleading conclusions. Faith in the robustness of least-squares methodology is no substitute for careful examination of the data using box-plots and histograms. Furthermore, although least-squares methods may provide valid estimators of mean cost, the estimators may be inefficient in the presence of right skewing.

2.2.2   Using multiple regression models

In the presence of high levels of skewness, it is natural to consider whether, by modeling cost on covariates, the skewness in the data can be explained. It is helpful to recognize that, at the simplest level, the simple comparison of means from Section 2.2.1 can be represented by the following linear model

$$C_i = \alpha + \Delta_c t_i + \varepsilon_i$$

where α is an intercept term, t_i a treatment dummy, taking the values zero for the standard treatment (S) and one for the new treatment (T), and ε a random error term. The coefficient Δ_c on the treatment dummy gives the estimated incremental cost of treatment exactly as in Equation (2.1). Similarly, the standard error of the coefficient will be the square root of the variance given in Equation (2.2).

The advantage of the regression framework is that it is straightforward to add covariates in addition to the treatment indicator. For example, baseline patient characteristics measured prior to randomization can be employed to make allowance for prognostic information in the treatment comparison using the multiple regression framework given by

$$C_i = \alpha + \sum_{k=1}^{p} \beta_k x_{ik} + \Delta_c t_i + \varepsilon_i,$$

where x_{ik} is value of the kth covariate for the ith patient. As before, the coefficient Δ_c for the treatment indicator gives the incremental cost controlling for the covariates. In the context of an experimental design such as a randomized controlled trial, the randomization process is expected to ensure a balance of both observed and unobserved potentially confounding factors across the treatment arms. Therefore, the use of prognostic covariates will not usually materially affect the magnitude of the estimated incremental cost, but may account for some of the skewness and improve the precision of the estimate, leading to narrower confidence intervals, see Altman (1985) and Pocock (1984).

2.2.3   Transformation (and the retransformation problem)

In practice, adding covariates to a cost regression rarely results in a model with a high explanatory power. The consequence is that residual skewness in the error term of the regression often remains and consideration can be given to transforming the cost with the aim of fitting a superior model. However, with any transformation of the data, it is important to recognize that health care policy decisions must be made concerning costs on the untransformed scale and so retransformation from the scale of estimation back to the original cost scale will be required. Consider fitting a linear model to some transformation of the cost data $Z_i = g(C_i)$. Although the back-transformation $h(\cdot) = g^{-1}(\cdot)$ can be employed it is well-known that $E[h(\cdot)] \neq h(E[\cdot])$

for non-linear transformations, therefore estimating coefficients from the regression model

$$Z_i = \alpha^Z + \Delta_c^Z t_i + \varepsilon_i$$

in order to predict the cost on the untransformed scale as

$$\hat{C}_i = h\left(\hat{\alpha}^Z + \Delta_c^Z t_i\right)$$

would give a biased estimate since

$$E\left[h\left(Z_i\right)\right] = E\left[C_i\right]$$
$$= E\left[h\left(\alpha^Z + \Delta_c t_i + \varepsilon_i\right)\right]$$
$$\neq E\left[h\left(\alpha^Z + \Delta_c t_i\right)\right]$$

2.2.3.1   A Taylor series approximation

Commonly, the Taylor series (or Delta) method is used to provide an approximation of the expectation of a random variable under a non-linear transformation. Letting $\mu_Z = E[Z]$, we have:

$$E\left[h\left(Z\right)\right] = E\left[h\left(\mu_Z + Z - \mu_Z\right)\right]$$
$$= E\left[h\left(\mu_Z\right) + \left(Z - \mu_Z\right)h'\left(\mu_Z\right) + \frac{\left(Z - \mu_Z\right)^2}{2!}h''\left(\mu_Z\right)\right.$$
$$\left. + \frac{\left(Z - \mu_Z\right)^3}{3!}h'''\left(\mu_Z\right) + \ldots\right] \simeq h\left(\mu_Z\right) + \frac{1}{2}h''\left(\mu_Z\right)\operatorname{var}\left[Z\right]$$

Thus the Taylor series approximation to the expectation on the untransformed scale suggests using a bias correction term of one half of the second derivative of the back-transformation function $h(\cdot)$ multiplied by the variance of Z, which can be estimated by $\operatorname{var}\left(\hat{\varepsilon}_i^Z\right)$.

2.2.3.2   A non-parametric 'smearing' estimator

While the Taylor series approximation may be adequate, there is an alternative approach known as non-parametric smearing that does not require a specific distribution for the error term, see Duan (1983). Note that the expectation of the untransformed cost can be written as a functional of the cumulative distribution function (cdf) of the errors on the transformed scale

$$E\left[C_i\right] = E\left[h\left(\alpha^Z + \Delta_c^Z t_i + \varepsilon_i^Z\right)\right]$$
$$= \int h\left(\alpha^Z + \Delta_c^Z t_i + \varepsilon_i^Z\right) dF\left(\varepsilon_i^Z\right),$$

where $F\left(\varepsilon_i^Z\right)$ represents the cumulative distribution function (cdf) of the error on the transformed scale. Rather than assume a parametric form for the cdf, it can be estimated by the empirical cdf of the estimated residuals

$$\hat{F}_t\left(e\right) = \frac{1}{n}\sum_{i=1}^{n} I\left\{\hat{\varepsilon}_i^z \leq e\right\},$$

where $I\left\{\cdot\right\}$ is the indicator function, $\hat{\varepsilon}_i^Z = Z_i - \left(\hat{\alpha}^Z + \hat{\Delta}_c^Z t_i\right)$, $Z_i = g(C_i)$, $n = n_S + n_T$, and $\hat{\alpha}^Z$ and $\hat{\Delta}_c^Z$ are the OLS estimated regression coefficients of Z_i on t_i.

The expected cost $E\left[C_i\right]$ is estimated by substituting the empirical cdf for the unknown cdf and substituting the OLS estimates of the regression coefficients to yield

$$\hat{E}\left[C_i\right] = \int h\left(\hat{\alpha}^Z + \hat{\Delta}_c^Z t_i + \hat{\varepsilon}_i^Z\right) d\hat{F}_t\left(\hat{\varepsilon}_i^Z\right) = \frac{1}{n}\sum_{i=1}^{n} h\left(\hat{\alpha}^Z + \hat{\Delta}_c^Z t_i + \hat{\varepsilon}_i^Z\right),$$

which gives the 'smearing' estimate for the expectation on the untransformed scale and provides an estimate of the between-treatment difference given by

$$\hat{\Delta}_c = \frac{1}{n}\sum_{i=1}^{n} h\left(\hat{\alpha}^Z + \hat{\Delta}_c^Z + \hat{\varepsilon}_i^Z\right) - \frac{1}{n}\sum_{i=1}^{n} h\left(\hat{\alpha}^Z + \hat{\varepsilon}_i^Z\right).$$

2.2.4   Generalized linear models

An alternative approach to transforming the data is to work within the class of generalized linear models (GLMs) where a linear predictor, $x^T\beta$, is related to the expectation of the outcome of interest through a link function $g(\cdot)$, such that $g\left(E\left[y|x\right]\right) = x^T\beta$. Since, in the GLM framework, it is the expectation that is subject to the transformation (rather than the data), back-transformation to the original scale is straightforward using the inverse of the link function: $E[y|x] = g^{-1}(x^T \beta)$. Given the fundamental interest of the economist in the mean cost/expenditure, this makes the GLM class of models particularly attractive. In addition, GLMs are extremely flexible, allowing a range of different distributions for the data to be coupled with different link functions.

For the purposes of the exposition here, the multiplicative log link GLM is assumed such that the model for cost is given by

$$\ln(E[C_i]) = \alpha + \sum_{k=1}^{p} \beta_k x_{ik} + \Delta_c t_i$$

or, equivalently

$$E[C_i] = \exp\left\{\alpha + \sum_{j=1}^{p} \beta_j x_{ij} + \Delta_c t_i\right\}$$

$$= \exp\left\{\alpha + \sum_{k=1}^{p} \beta_k x_{ik}\right\} \cdot \exp\{\Delta_c t_i\}$$

(Model 4)

What should be clear for the expression given above, is that the coefficient on the treatment dummy from a multiplicative model of cost is a factor, such that the incremental cost is obtained by multiplying this factor by the average cost in the absence of treatment. This means that the estimated incremental cost will vary according to the baseline cost in the standard care arm. If covariates have an important impact on this baseline cost, then the incremental cost of treatment may also vary substantially by covariate pattern. Thus, it is the lack of treatment interaction with baseline covariates in a multiplicative model that indicates important subgroup effects.

2.2.5   Two-part models for excess zeros

One approach to the problem of excess zeros is to form a two-part model. In the first part of the model, a logistic GLM is employed using the indicator variable for positive cost to predict which costs will be positive and which costs will be zero. This gives the first part of the model as

$$\ln\left(\frac{\pi_i}{1-\pi_i}\right) = \alpha^\pi + \Delta^\pi t_i$$

with the predicted probability that the cost is positive given by

$$\pi_i = \frac{\exp\{\alpha^\pi + \Delta^\pi t_i\}}{1 + \exp\{\alpha^\pi + \Delta^\pi t_i\}}$$

The second part of the model then involves modeling the positive costs only using the transformed positive costs C_i^+, the transformed positive costs $Z_i^+ = g\left(C_i^+\right)$ using OLS, or most flexibly with a GLM for the expectation of positive cost.

The overall predicted value for a two-part model is simply the product of the expectations of its two parts, given by

$$\hat{C}_i = \hat{\pi}_i \hat{C}_i^+ = \frac{\exp\{\hat{\alpha}^\pi + \hat{\Delta}^\pi t_i\}}{1 + \exp\{\hat{\alpha}^\pi + \hat{\Delta}^\pi t_i\}} g^{-1}\left(\hat{\alpha}^{C^+} + \hat{\Delta}^{C^+} t_i\right)$$

where the regression of positive cost is assumed undertaken from a GLM with $g^{-1}(\cdot)$ representing the inverse of the link function. The difference between treatment arms is estimated by

$$\frac{\exp\{\hat{\alpha}^\pi + \hat{\Delta}^\pi\}}{1 + \exp\{\hat{\alpha}^\pi + \hat{\Delta}^\pi\}} g^{-1}\left(\hat{\alpha}^{C^+} + \hat{\Delta}^{C^+}\right) - \frac{\exp\{\hat{\alpha}^\pi\}}{1 + \exp\{\hat{\alpha}^\pi\}} g^{-1}\left(\hat{\alpha}^{C^+}\right)$$

A similar result holds if the second part of the two-part model is based on transformed cost; however, as discussed above, consideration should be given to a Taylor series or smearing correction in order to give an unbiased estimators for untransformed cost.

2.2.6   Cost prediction models

Collecting all health care utilization data in a clinical trial can be very expensive. As an alternative, investigators may choose to collect a small number of high-impact utilization variables, and use them, along with some baseline and outcome variables, to predict total cost for each patient. This approach relies on the existence of the results of a regression analysis using a previously existing dataset, in which total cost has been regressed on the predictor variables in question.

Let x_{jki} be the observed value of predictor k ($k = 1, \ldots p$) on patient i ($i = 1, \ldots n_j$) from treatment arm j ($j = T, S$). Then $x_{ji} = (x_{j1i}, x_{j2i}, \ldots x_{jpi})^T$ is the vector of predictor variables for patient i from treatment arm j. Let $\bar{x}_j = \frac{1}{n_j} \Sigma_{i=1}^{n_j} x_{ji}$, $\bar{x} = \bar{x}_T - \bar{x}_S$, $\hat{\Sigma}_{xj}$ be the sample variance–covariance matrix based on the x_{ji} values, and $\hat{\Sigma}_x = \frac{1}{n_T} \hat{\Sigma}_{xT} + \frac{1}{n_S} \hat{\Sigma}_{xS}$. Let $\hat{\beta}$ be the vector of dimension p of parameter estimates corresponding to the p predictor variables resulting from the regression analysis performed on the existing data, with estimated variance–covariance matrix given by $\hat{\Sigma}_\beta$. The quantity $\hat{\beta}_0 + \hat{C}_{ji}$ is the predicted cost for patient i from treatment arm j, where $\hat{C}_{ji} = x_{ji}^T \hat{\beta}$ and $\hat{\beta}_0$ is the estimate of the intercept from the regression analysis performed on the existing data. The estimator of between-arm difference in mean cost is $\bar{\hat{C}}_T - \bar{\hat{C}}_S$, where $\bar{\hat{C}}_j = \frac{1}{n_j} \Sigma_{i=1}^{n_j} \hat{C}_{ji}$. A naïve estimator for the variance of $\bar{\hat{C}}_T - \bar{\hat{C}}_S$ is given by

$$\frac{\sum\limits_{i=1}^{n_S} (\hat{C}_{Si} - \bar{\hat{C}}_S)^2}{n_S (n_S - 1)} + \frac{\sum\limits_{i=1}^{n_T} (\hat{C}_{Ti} - \bar{\hat{C}}_T)^2}{n_T (n_T - 1)}$$

but this ignores the fact that $\hat{\beta}$ is a random vector. The proper estimator for the variance of $\bar{\hat{C}}_T - \bar{\hat{C}}_S$ is given by $x^T \hat{\Sigma}_\beta x + \hat{\beta}^T \hat{\Sigma}_x \hat{\beta} + \text{trace} (\hat{\Sigma}_\beta \hat{\Sigma}_x)$, (see Willan and O'Brien, 2001). Setting $\hat{\Sigma}_\beta = 0$ yields the naïve estimator given above.

2.3   EFFECTIVENESS

In dealing with the effectiveness side of cost-effectiveness analysis, consideration is given to each of the three possible measures of effectiveness typically employed in clinical trials that include economic analyses: probability of surviving, mean survival time and mean quality adjusted survival time. Due to the role of baseline utility measures in the QALY calculus, the appropriate adjustment of QALY measures is also discussed.

2.3.1 Probability of surviving

Let the random variable D_{ji} be the time from randomization to death for the jith patient, and let $S_j(t) = \Pr(D_{ji} \geq t)$, then the probability of surviving the duration of interest for a patient on the jth arm is given by $S_j(\tau)$, which is denoted more simply by π_j. For non-censored data $S_j(\tau)$ is estimated by the proportion of patients who survived the duration of interest. Let $\bar{\delta}_{ji} = 1$ if the jith patient was observed to survive the duration of interest, and 0 if not. That is, $\bar{\delta}_{ji} = I\left\{D_{ji} \geq \tau\right\}$, where $I\{\text{expression}\}$ is the indicator function, equaling 1 if the expression is true, and 0 otherwise. The probability of surviving the duration of interest for a patient on the jth arm is estimated by $\hat{\pi}_j = \frac{1}{n_j}\sum_{i=1}^{n_j} \bar{\delta}_{ji}$. Therefore, the estimated between-arm difference is given by

$$\hat{\Delta}_e = \hat{\pi}_T - \hat{\pi}_S = \frac{1}{n_T}\sum_{i=1}^{n_T} \bar{\delta}_{Ti} - \frac{1}{n_S}\sum_{i=1}^{n_S} \bar{\delta}_{Si} \qquad (2.3)$$

The variance of $\hat{\Delta}_e$ is estimated by

$$\hat{V}\left(\hat{\Delta}_e\right) = \hat{V}\left(\hat{\pi}_T\right) + \hat{V}\left(\hat{\pi}_S\right) = \frac{\hat{\pi}_T(1-\hat{\pi}_T)}{n_T} + \frac{\hat{\pi}_S(1-\hat{\pi}_S)}{n_S} \qquad (2.4)$$

and the covariance between $\hat{\Delta}_e$ and $\hat{\Delta}_c$ by

$$\hat{\text{Cov}}\left(\hat{\Delta}_e, \hat{\Delta}_c\right) = \hat{\text{Cov}}\left(\hat{\pi}_T, \hat{v}_T\right) + \hat{\text{Cov}}\left(\hat{\pi}_S, \hat{v}_S\right)$$

$$= \frac{\left[\left(\sum_{i=1}^{n_T} \bar{\delta}_{Ti}C_{Ti}\right) - n_T\hat{\pi}_T\hat{v}_T\right]}{n_T(n_T-1)} \qquad (2.5)$$

$$+ \frac{\left[\left(\sum_{i=1}^{n_S} \bar{\delta}_{Si}C_{Si}\right) - n_S\hat{\pi}_S\hat{v}_S\right]}{n_S(n_S-1)}$$

2.3.2 Mean survival time

Since there is no censoring, all patients have their survival time observed and sample means can be used to estimate the mean survival time. Let $X_{ji} = \min(D_{ji},\tau)$; i.e. X_{ji} is the duration of interest or the time from randomization to death, whichever is smaller. The mean survival for a patient on the jth treatment, denoted by μ_j, is estimated by: $\hat{\mu}_j = \frac{1}{n_j}\Sigma_{i=1}^{n_j} X_{ji}$. Therefore, the estimated between-arm difference is given by

$$\hat{\Delta}_e = \hat{\mu}_T - \hat{\mu}_S = \frac{1}{n_T} \sum_{i=1}^{n_T} X_{Ti} - \frac{1}{n_S} \sum_{i=1}^{n_S} X_{Si} \qquad (2.6)$$

The variance of $\hat{\Delta}_e$ is estimated by

$$\hat{V}\left(\hat{\Delta}_e\right) = \hat{V}\left(\hat{\mu}_T\right) + \hat{V}\left(\hat{\mu}_S\right)$$

$$= \frac{1}{n_T(n_T - 1)} \sum_{i=1}^{n_T} (X_{Ti} - \hat{\mu}_T)^2 + \frac{1}{n_S(n_S - 1)} \sum_{i=1}^{n_S} (X_{Si} - \hat{\mu}_S)^2$$

$$(2.7)$$

and the covariance between $\hat{\Delta}_e$ and $\hat{\Delta}_c$ by

$$\hat{Cov}\left(\hat{\Delta}_e, \hat{\Delta}_c\right) = \hat{Cov}\left(\hat{\mu}_T, \hat{v}_T\right) + \hat{Cov}\left(\hat{\mu}_S, \hat{v}_S\right)$$

$$= \frac{\sum_{i=1}^{n_T} (X_{Ti} - \hat{\mu}_T)(C_{Ti} - \hat{v}_T)}{n_T(n_T - 1)} \qquad (2.8)$$

$$+ \frac{\sum_{i=1}^{n_S} (X_{Si} - \hat{\mu}_S)(C_{Si} - \hat{v}_S)}{n_S(n_S - 1)}$$

2.3.3   Mean quality-adjusted survival time

To measure a patient's quality-adjusted life-years (QALYs) over the duration of interest, we assume that the patients quality of life (QoL) is measured at various times during this period, most likely, though not necessarily at $t = 0$ (at randomization), at $t = \tau$, and perhaps at times in between. The measurements are on a scale in which 1 corresponds to perfect health, 0 to death, and negative values to states of health worse than death. For a more complete discuss of measuring quality of life, the reader is referred to Weinstein and Stason (1977) and Torrance (1976; 1986; 1987).

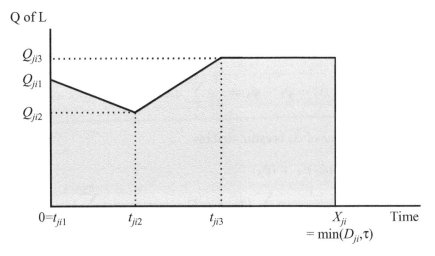

Figure 2.1 Quality of life by time

Suppose there are m_{ji} QoL measurements on the jith patient, taken at times: $0 \le t_{ji1} < t_{ji2} < \ldots < t_{jim_{ji}} \le \tau$, with corresponding values $Q_{ji1}, Q_{ji2}, \ldots Q_{jim_{ji}}$. The observed QALY on the jith patient, denoted by q_{ij}, is simply the area under the curve, between 0 and X_{ji}, of the plot of the Q_{ij} by the t_{ij} values, as shown in Figure 2.1. (As would be expected with most clinical trials, in Figure 2.1 we have set $t_{ji1} = 0$.) Therefore, $q_{ji} = \int_0^{X_{ji}} Q_{ji}(t)dt$, where $Q_{ji}(t)$ is defined as

$$
Q_{ji}(t) = \begin{cases}
Q_{ji1} : & 0 \le t < t_{ji1} \\
Q_{jih} + \dfrac{(Q_{ji,h+1} - Q_{jih})(t - t_{jih})}{t_{ji,h+1} - t_{jih}} : & t_{jih} \le t < t_{ji,h+1} \\
Q_{jim_{ji}} : & t_{jim_{ji}} \le t < X_{ji} \\
0 : & t \ge X_{ji}
\end{cases} \tag{2.9}
$$

More simply put, q_{ij} is just the sum of the areas of the shaded rectangles shown in Figure 2.1, i.e. $q_{ji} = (t_{ji2} - t_{ji1}) \times \left(\frac{Q_{ji1}+Q_{ji2}}{2}\right) + (t_{ji3} - t_{ji2}) \times \left(\frac{Q_{ji2}+Q_{ji3}}{2}\right) + (X_{ji} - t_{ji3}) \times Q_{ji3}$. If t_{ji1} is greater than 0, the definition in Equation (2.9) sets the QoL between 0 and t_{ji1} to a constant value of Q_{ji1}. Similarly, if X_{ji} is greater than $t_{jim_{ji}}$ then Equation (2.9) sets the QoL between $t_{jim_{ji}}$ and X_{ji} to a constant value of $Q_{jim_{ji}}$.

The mean QALY for a patient on the jth arm, denoted by φ_j, is estimated by: $\hat{\varphi}_j = \frac{1}{n_j} \sum_{i=1}^{n_j} q_{ji}$. Therefore, the estimated between-arm difference is given by:

$$\hat{\Delta}_e = \hat{\varphi}_T - \hat{\varphi}_S = \frac{1}{n_T} \sum_{i=1}^{n_T} q_{Ti} - \frac{1}{n_S} \sum_{i=1}^{n_S} q_{Si} \qquad (2.10)$$

The variance of $\hat{\Delta}_e$ is estimated by

$$\hat{V}\left(\hat{\Delta}_e\right) = \hat{V}\left(\hat{\varphi}_T\right) + \hat{V}\left(\hat{\varphi}_S\right)$$

$$= \frac{1}{n_T(n_T - 1)} \sum_{i=1}^{n_T} (q_{Ti} - \hat{\varphi}_T)^2 + \frac{1}{n_S(n_S - 1)} \sum_{i=1}^{n_S} (q_{Si} - \hat{\varphi}_S)^2$$

$$(2.11)$$

and the covariance between $\hat{\Delta}_e$ and $\hat{\Delta}_c$ by

$$\hat{Cov}\left(\hat{\Delta}_e, \hat{\Delta}_c\right) = \hat{Cov}\left(\hat{\varphi}_T, \hat{v}_T\right) + \hat{Cov}\left(\hat{\varphi}_S, \hat{v}_S\right)$$

$$= \frac{\sum_{i=1}^{n_T} (q_{Ti} - \hat{\varphi}_T)(C_{Ti} - \hat{v}_T)}{n_T(n_T - 1)} + \frac{\sum_{i=1}^{n_S} (q_{Si} - \hat{\varphi}_S)(C_{Si} - \hat{v}_S)}{n_S(n_S - 1)}$$

$$(2.12)$$

2.3.4   Mean quality-adjusted survival time: controlling for baseline utility

In principle, the methods of Section 2.2.2 which described the use of regression models for cost apply equally to effectiveness outcomes, in that even where perfect balance is achieved in the randomization process, the inclusion of additional covariates into a regression framework can be expected to reduce the variance and improve the power of the analysis. However, in terms of QALY outcomes there is an important distinction to be made between baseline measures of covariates that, while prognostic for outcomes, are independent of treatment (through the randomization process) and baseline measures of utility, which forms part of the QALY calculus. Being part of the QALY calculation, and therefore correlated with the QALY outcomes, any imbalance in baseline utility measures will potentially bias estimates of $\hat{\Delta}_e$ above.

In order to adjust appropriately for potential imbalance, the baseline utility can be added to the multiple regression equation to give the QALY estimate as

$$q_i = \alpha + \sum_{k=1}^{p} \beta_k x_{ik} + \Delta_e t_i + \gamma Q_i(0) + \varepsilon_i$$

where t_i is a treatment indicator variable as before, $Q_i(0)$ is the baseline utility score, and the x_{ij} are other potential prognostic covariates (as in Section 2.2.2). Further discussion on the use of regression adjustment for handling imbalance at baseline can be found in Altman (1985) and Senn (1997). Manca *et al.* (2005a) discuss the baseline adjustment of QALYs in particular and present simulation results to show that regression adjustment outperforms simple measures of change from baseline, which is sometimes used as a method for adjusting for baseline imbalance in the outcome of interest.

2.4 SUMMARY

Methods for estimating the five parameters required for a cost-effectiveness comparison of two groups using non-censored data were discussed in this chapter. The main focus was on the simple comparison of two arms of a clinical trial, but emphasizing the potential important of regression methods for increasing the precision of estimates of treatment effect, adjusting for imbalance, and avoiding bias associated with baseline utility imbalances affecting the QALY calculations. The potential treatment of skewness and excess zeros in the analysis of cost data were also discussed, with an emphasis on the use of GLMs to focus on the fundamental interest on mean values. The utilization of these five parameters in a cost-effectiveness analysis is the subject of Chapter 4, while the use of covariates in a cost-effectiveness analysis and the examination of sub-group effects is the subject of Chapter 7.

3

Parameter Estimation for Censored Data

3.1 INTRODUCTION

The analysis of a clinical trial involves censored data if some patients are not followed for the entire duration of interest. Some patients may be lost to follow-up, either because they refuse to attend follow-up clinic visits or because they move out of the jurisdiction covered by trial management resources. Because of staggered entry, patients also can be censored because the analysis is performed prior to them being followed for the duration of interest. This type of censoring, know as administrative, is present more frequently for trials with relatively long durations of interest, say 3–5 years.

Lost-to-follow-up (LTF) censoring is usually considered more problematic, since it is harder to assume that the distributions for the outcomes (cost and effectiveness) for LTF patients are the same as those not lost to follow-up. If the survival experience of those LTF does differ, then the censoring is informative, leading to biased parameter estimators. This potential source of bias is exacerbated if LTF rates or the causes of LTF differ between treatment arms. Administrative censoring is less problematic than FTF censoring if it can be assumed that patients randomized later in the trial, and therefore more likely to be censored administratively, have the same survival and cost distributions as those randomized earlier. This may not always be true, since co-interventions that improve survival or alter costs may become

Statistical Analysis of Cost-effectiveness Data. A. Willan and A. Briggs
© 2006 John Wiley & Sons, Ltd.

available during the recruitment phase of the trial, improving the survival times or altering the costs for patients randomized later. Also, entry criteria are often changed during the trial, and consequently, survival and cost distributions may depend on when patients were randomized. Nonetheless, most trials with censored data tend to assume that censoring is uninformative, or more accurately, assume that the extent of the informative censored is unlikely to bias the estimates of treatment effect appreciably.

Parameter estimators for censored data are given in the remaining sections of this chapter. Standard life-table methods are used for probability of surviving and the mean survival time. Two methods are proposed for mean cost and mean quality-adjusted survival time. The direct (Lin) method makes use of the survival function combined with the cost histories, while the method of inverse-probability weighting makes use of the censoring function together with the cost histories. Inverse-probability weighting can also be used for estimating mean survival time.

Recall that D_{ji} is the time from randomization to death for the jith patient. Let U_{ji} be the time from randomization to censoring for the jith patient. The jith patient is censored if $U_{ji} \leq D_{ji}$ and is observed to die if $D_{ji} < U_{ji}$. Further, let $X_{ji} = \min(D_{ji}, U_{ji})$, $\delta_{ji} = I\{D_{ji} < U_{ji}\}$ and $\bar{\delta}_{ji} = 1 - \delta_{ji}$. Therefore X_{ji} is the time on study, δ_{ji} is the indicator for observed death, and $\bar{\delta}_{ji}$ is the indicator for censoring.

3.2 MEAN COST

Because patients accrue costs at different rates, the cost accrued to death and the cost accrued to censoring will be positively correlated, even though the times to death and censoring are not. A patient accruing cost at a fast rate will tend to have a large cost at death and a large cost at censoring; and conversely, a patient accruing cost at a slow rate will tend to have a small cost at death and a small cost at censoring. As a result informative censoring is induced on the cost scale, and the use of standard life-table methods for estimating mean cost, similar to those for estimating mean survival time (see Section 3.3.2.1), will lead to positively biased estimators. Because patients with smaller cost are more likely to be censored, the estimator of mean cost will be positively biased, providing an overestimate of the mean cost. Two methods have been proposed to account for this induced informative censoring on the cost scale, the direct (Lin) method and inverse-probability weighting.

3.2.1   Direct (Lin) method

This method was first proposed by Lin *et al.* (1997), and requires that the duration of interest be divided into K intervals $[a_k, a_{k+1})$, where $0 = a_1 < a_2 < \ldots < a_{K+1} = \tau$, in which the cost of each patient is known, at least until the patient is censored. The set of costs incurred during these intervals is often referred to as the cost histories. The interval $[a_k, a_{k+1})$ includes all those values of t such that $a_k \leq t < a_{k+1}$. These intervals, which need not be of equal length, usually coincide with the follow-up clinic visits when health care utilization data are collected.

Let C_{jki} be the cost of the *ji*th patient during the kth interval. If the *ji*th patient is censored during the kth interval, then C_{jki} is unknown for the kth and subsequent intervals. If the *ji*th patient dies during the kth interval, then C_{jki} is zero for all subsequent intervals. The assumption is made here that the cost for the interval in which a patient dies is known. Let $\delta_{jki}^{**} = 1$ if the *ji*th patient enters the kth interval (i.e. $X_{ji} > a_k$) and either dies ($\delta_{ji} = 1$) or is not censored during the interval ($X_{ji} > a_{k+1}$); otherwise $\delta_{jki}^{**} = 0$. The value of δ_{jki}^{**} is 1 for all patients on whom complete cost data for the kth interval is observed. The value of δ_{jki}^{**} is 0 for all patients who either died prior to a_k or were censored prior to a_{k+1}, i.e. those patients for whom cost data was not observed during the kth interval. Algebraically, $\delta_{jki}^{**} = I\left\{X_{ji} > a_k \text{ and } \left(X_{ji} > a_{k+1} \text{ or } \delta_{ji} = 1\right)\right\}$. Let $\tilde{C}_{jk}$ be the average of the cost of the patients who have complete cost data observed in the kth interval, i.e. those for whom $\delta_{jki}^{**} = 1$. Thus $\tilde{C}_{jk} = (\sum_{i=1}^{n_j}\delta_{jki}^{**})^{-1}\sum_{i=1}^{n_j}\delta_{jki}^{**}C_{jki}$. Assuming that the censored data in the kth interval are missing completely at random, $\tilde{C}_{jk}$ provides an unbiased estimator of the mean cost during the kth interval of patients on the jth arm who survive beyond a_k. The estimator of v_j, the mean cost over the duration of interest, is simply the sum of the $\tilde{C}_{jk}$, weighted by the probability of surviving to the beginning of the interval. That is,

$$\hat{v}_j = \sum_{k=1}^{K} S_j(a_k) \tilde{C}_{jk} \qquad (3.1)$$

The estimator is unbiased if the censoring occurs only at the boundaries of the intervals. In practice the survival function $S_j(\cdot)$ must be estimated, and so Equation (3.1) becomes

$$\hat{v}_j = \sum_{k=1}^{K} \hat{S}_j(a_k) \tilde{C}_{jk}. \qquad (3.2)$$

The estimator in Equation (3.2) is consistent if the censoring occurs only at the boundaries of the intervals.

The survival function can be estimated using the product-limit method. Let $t_{j1}, t_{j2}, t_{j3}, \ldots$ be the unique times at which deaths occur on the jth arm, let $d_{j1}, d_{j2}, d_{j3}, \ldots$ be the number of deaths occurring at those times, and let $n_{j1}, n_{j2}, n_{j3}, \ldots$ be the number at risk of dying (i.e. not previously dead or censored) at those times. Then the estimator of survival function is given by

$$\hat{S}_j(t) = \prod_{h | t_{jh} < t} \left(1 - \frac{d_{jh}}{n_{jh}} \right) \qquad \text{for } t_{j,h-1} \leq t < t_{jh} \qquad (3.3)$$

$\hat{S}_j(\cdot)$ is a step function that decreases at those times at which deaths occur, see Figure 3.1. The height of $\hat{S}_j(\cdot)$ at the beginning of the interval provides the weight in the sum of the $\tilde{C}_{jk}$ values. The estimator of the between-arm difference in mean cost is given by $\hat{\Delta}_c = \hat{v}_T - \hat{v}_S$.

The estimator of the variance of $\hat{v}_j$ is given by

$$\hat{V}\left(\hat{v}_j\right) = \sum_{i=1}^{n_j} \sum_{k=1}^{K} \sum_{g=1}^{K} \hat{\xi}_{jki}^{(c)} \hat{\xi}_{jgi}^{(c)} \qquad (3.5)$$

where

$$\hat{\xi}_{jki}^{(c)} = \left(\sum_{i=1}^{n_j} \delta_{jki}^{**} \right)^{-1} \left(C_{ijk} - \tilde{C}_{jk} \right) \hat{S}_j(a_k) \delta_{jki}^{**}$$
$$- \hat{S}_j(a_k) \tilde{C}_{jk} \left(\frac{I\{X_{ji} \leq a_k\} \delta_{ji}}{R_{ji}} - \sum_{g=1}^{n_j} \frac{I\{X_{jg} \leq \min(X_{ji}, a_k)\} \delta_{jg}}{R_{jg}^2} \right)$$

$$(3.6)$$

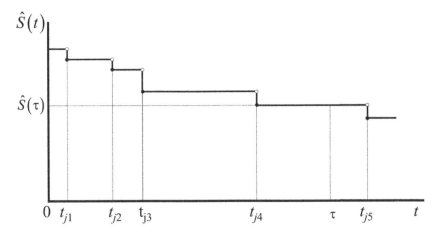

Figure 3.1 Estimated survival curve

and $R_{ji} = \sum_{g=1}^{n_j} I \left\{ X_{jg} \geq X_{ji} \right\}$, see Lin *et al.* (1997) and Willan and Lin (2001). Therefore the estimator of the variance of $\hat{\Delta}_c$ is given by

$$\hat{V}\left(\hat{\Delta}_c\right) = \hat{V}\left(\hat{v}_T\right) + \hat{V}\left(\hat{v}_S\right) = \sum_{i=1}^{n_T}\sum_{k=1}^{K}\sum_{g=1}^{K}\hat{\xi}_{Tki}^{(c)}\hat{\xi}_{Tgi}^{(c)} + \sum_{i=1}^{n_S}\sum_{k=1}^{K}\sum_{g=1}^{K}\hat{\xi}_{Ski}^{(c)}\hat{\xi}_{Sgi}^{(c)}$$

$$(3.7)$$

If cost histories are not available and only total cost for the duration of interest is observed, then there is a single interval $[0, \tau)$, and $\hat{v}_j$ is the sample mean for the non-censored patients and $\hat{V}\left(\hat{v}_j\right)$ is the sample variance for the non-censored patients, divided by the number of non-censored patients.

3.2.2   Inverse-probability weighting

The method of inverse-probability weighting (IPW) can be used as an alternative to the direct method for estimating mean cost in the presence of censoring, see Bang and Tsiatis (2000), Lin (2000), Zhao and Tian (2001) and Willan *et al.* (2002). The principle behind IPW can be illustrated by the following simple example.

Let the random variables Y_i, $i = 1, 2, \ldots N$ be normally distributed (μ, σ^2), and let $\delta_i = I\{Y_i \text{ is observed}\}$, i.e. δ_i is 0 if Y_i is censored. Let n be the number of non-censored observations, i.e. $n = \Sigma_{i=1}^N \delta_i$. If we assume that the censored data is missing at random, then μ is estimated by the average of the non-censored observations, and can be written, rather clumsily, as

$$\hat{\mu} = \left(\sum_{i=1}^N \delta_i \right)^{-1} \sum_{i=1}^N \delta_i Y_i = \left(\sum_{i=1}^N \frac{\delta_i}{n/N} \right)^{-1} \sum_{i=1}^N \frac{\delta_i Y_i}{n/N} \tag{3.8}$$

The last equality holds because the numerator and denominator are both being divided by n/N, albeit unnecessarily. Thus, $\hat{\mu}$ is a weighted average of the Y_i values, where the weight for each observation is δ_i divided by the n/N, i.e. δ_i divided by the probability of not being censored. The factor n/N is not really required in this example because the probability of not being censored is the same for all observations. However, in a follow-up study the probability of an observation not being censored will depend on when that observation is taken.

We will apply IPW to provide an estimator of mean cost in the presence of censoring. Let v_{jk} be the mean cost for a patient on the jth arm during the kth interval. Therefore, $v_j = \Sigma_{i=1}^K v_{jk}$. We use IPW to provide consistent estimators of the v_{jk}, which can be summed to provide a consistent estimator of v_j. Let $G_j(t) = \Pr(U_{ji} \geq t)$, i.e. $G_j(t)$ is the probability of a patient on the jth treatment not being censored by time t. Let $\overset{*}{\delta}_{jki}$ equal to 1 if the jith patient dies (i.e. $\delta_{ji} = 1$) or is not censored before a_{k+1} (i.e. $X_{ji} \geq a_{k+1}$). For a patient who dies, $\overset{*}{\delta}_{jki}$ is equal to 1 for all intervals, including those following his or her death. For a patient who is censored, $\overset{*}{\delta}_{jki}$ is equal to 1 for intervals prior to the censoring, and equal to 0 for the interval in which the censoring occurs and all subsequent intervals. Algebraically,

$$\overset{*}{\delta}_{jki} = \delta_{ji} + \bar{\delta}_{ji} I \left\{ X_{ji} \geq a_{k+1} \right\} \tag{3.9}$$

Further, let $\overset{*}{X}_{jki} = \min \left(X_{ji}, a_{k+1} \right)$. The value of $\overset{*}{X}_{jki}$ is equal to the upper boundary of the interval for those intervals prior to the patient's death or censoring, and equal to the time on study, i.e. X_{ji}, for the interval in which the death or censoring occurs and for all subsequent intervals. An unbiased estimator of v_{jk} is given by

$$\hat{v}_{jk} = \left(\sum_{i=1}^{n_j} \frac{\delta_{jki}^*}{G\left(X_{jki}^*\right)} \right)^{-1} \sum_{i=1}^{n_j} \frac{\delta_{jki}^* C_{jki}}{G\left(X_{jki}^*\right)} \tag{3.10}$$

Equation (3.10) is analogous to Equation (3.9). The value of δ_{jki}^* is 1 if C_{jki} is observed (recall that the cost during an interval following death is set to 0, since it is assumed that patients stop accumulating costs after death). The divisor $G\left(X_{jki}^*\right)$ is the probability of not being censored, evaluated at the end of the interval if the patient is on study until the end, or at the time of death if the patient dies. If a patient is censored during the interval, δ_{jki}^* is equal to 0, and the patient will not contribute to the estimated mean for that interval. The role of the divisor is to inflate the observed C_{jki} to account for the censored data. If there is no censoring $G\left(X_{jki}^*\right)$ equals 1 and no inflation occurs. If $G\left(X_{jki}^*\right)$ is 0.8, for sake of argument, then the value of C_{jki} is inflated by 1.25 (i.e. 1/0.8) to make up for the censored data when determining the total cost incurred by all patients, both censored and non-censored. The total is then divided by the sum of the weights to derive the mean.

It is usually necessary to use the product-limit to estimate $G_j(\cdot)$. Similarly to estimating $S_j(\cdot)$,

$$\hat{G}_j(t) = \prod_{h|t_{jh}^* < t} \left(1 - \frac{d_{jh}^*}{n_{jh}^*} \right) \quad \text{for } t_{jh-1}^* \leq t < t_{jh}^* \tag{3.11}$$

where the t_{jh}^*'s are the unique censoring times for the jth arm, and n_{jh}^* and d_{jh}^* are, respectively, the number of patients at risk for censoring and the number of patients censored at time t_{jh}^*. Equation (3.10) becomes

$$\hat{v}_{jk} = \left(\sum_{i=1}^{n_j} \frac{\delta_{jki}^*}{\hat{G}\left(X_{jki}^*\right)} \right)^{-1} \sum_{i=1}^{n_j} \frac{\delta_{jki}^* C_{jki}}{\hat{G}\left(X_{jki}^*\right)} \tag{3.12}$$

and is an estimator for the mean cost during interval k. The estimator for v_j is given by

$$\hat{v}_j = \sum_{k=1}^{K} \hat{v}_{jk} \tag{3.13}$$

The estimator given in Equation (3.13) is consistent, regardless of the pattern of censoring. This is in contrast to the estimator provided by the direct method, which is consistent only if the censoring occurs at the boundaries of the intervals.

The estimator of the between-arm difference in mean cost is given by $\hat{\Delta}_c = \hat{v}_T - \hat{v}_S$, and the variance of $\hat{\Delta}_c$ is given by

$$\hat{V}\left(\hat{\Delta}_c\right) = \hat{V}\left(\hat{v}_T\right) + \hat{V}\left(\hat{v}_S\right) = \sum_{i=1}^{n_T}\sum_{k=1}^{K}\sum_{g=1}^{K}\hat{\xi}_{Tki}^{(c)}\hat{\xi}_{Tgi}^{(c)} + \sum_{i=1}^{n_S}\sum_{k=1}^{K}\sum_{g=1}^{K}\hat{\xi}_{Ski}^{(c)}\hat{\xi}_{Sgi}^{(c)} \tag{3.14}$$

where

$$\hat{\xi}_{jki}^{(c)} = \frac{1}{n_j}\left(\frac{\overset{*}{\delta}_{jki}\left(C_{jki} - \hat{v}_{jk}\right)}{\hat{G}\left(X_{jki}^{*}\right)} + \bar{\delta}_{ji}B_{jki} - \sum_{g=1}^{n_j}\frac{\bar{\delta}_{jg}I\left\{X_{jg} \le X_{ji}\right\}B_{jkg}}{R_{jg}}\right) \tag{3.15}$$

and $B_{jki} = \frac{1}{R_{ji}}\Sigma_{g=1}^{n_j}\frac{I\{X_{jkg}^{*} > X_{ji}^{*}\}\delta_{jkg}(C_{jkg} - \hat{v}_{jk})}{\hat{G}(X_{jkg}^{*})}$, see Bang and Tsiatis (2000), Lin (2000), Zhao and Tian (2001) and Willan *et al.* (2002). To ease the burden of notation $\hat{\xi}_{jki}^{(c)}$ has two definitions, one in Equation (3.6) for the direct method and one in Equation (3.15), for IPW. This should not cause confusion since for any particular application only one method would be used.

3.3 EFFECTIVENESS

3.3.1 Probability of surviving

The probability of surviving the duration of interest can be estimated by $\hat{S}_j(\tau)$ and can be read from the estimated survival curve as shown in Figure 3.1. Formally, it is defined as

$$\hat{\pi}_j = \hat{S}_j(\tau) = \prod_{h \mid t_{jh} < \tau} \left(1 - \frac{d_{jh}}{n_{jh}}\right) \tag{3.16}$$

The estimator of the variance of $\hat{\pi}_j$ is given by $\hat{V}\left(\hat{\pi}_j\right) = \Sigma_{i=1}^{n_j}\left(\hat{\xi}_{ji}^{(p)}\right)^2$, where

$$\hat{\xi}_{ji}^{(p)} = -\hat{\pi}_j \left(\frac{I\{X_{ji} \leq \tau\}\delta_{ji}}{R_{ji}} - \sum_{g=1}^{n_j} \frac{I\{X_{jg} \leq \min(\tau, X_{ji})\}\delta_{jg}}{R_{jg}^2}\right) \tag{3.17}$$

see Willan *et al.* (2003). Therefore, when the measure of effectiveness is the probability of surviving the duration of interest, Δ_e is estimated by

$$\hat{\Delta}_e = \hat{\pi}_T - \hat{\pi}_S \tag{3.18}$$

and the variance of $\hat{\Delta}_e$ is given by

$$\hat{V}\left(\hat{\Delta}_e\right) = \hat{V}\left(\hat{\pi}_T\right) + \hat{V}\left(\hat{\pi}_S\right) = \sum_{i=1}^{n_T}\left(\hat{\xi}_{Ti}^{(p)}\right)^2 + \sum_{i=1}^{n_S}\left(\hat{\xi}_{Si}^{(p)}\right)^2 \tag{3.19}$$

The estimator of the covariance between $\hat{\Delta}_e$ and $\hat{\Delta}_c$ is given by

$$\hat{C}\left(\hat{\Delta}_e, \hat{\Delta}_c\right) = \hat{C}\left(\hat{\pi}_T, \hat{v}_T\right) + \hat{C}\left(\hat{\pi}_S, \hat{v}_S\right) = \sum_{i=1}^{n_T} \hat{\xi}_{Ti}^{(p)} \sum_{k=1}^{K} \hat{\xi}_{Tki}^{(c)}$$

$$+ \sum_{i=1}^{n_S} \hat{\xi}_{Si}^{(p)} \sum_{k=1}^{K} \hat{\xi}_{Ski}^{(c)} \tag{3.20}$$

see Willan *et al.* (2003). The definition of $\hat{\xi}_{jki}^{(c)}$ used in Equation (3.20) will depend on which method was used to estimate the mean cost. The appropriate definition is given in Equation (3.6) for the direct method and in Equation (3.15) for IPW.

3.3.2 Mean survival time

3.3.2.1 Area under the survival curve

The mean survival time over the duration of interest is the area under the survival curve from 0 to τ, i.e. $\int_0^\tau S_j(t)\,dt$, and can be estimated by the area under the estimated survival curve as shown in Figure 3.2. The estimator is simply the sum of the area of all the shaded rectangles, and, defining $t_{j0} = 0$, can be written algebraically as

$$\hat{\mu}_j = \int\limits_0^\tau \hat{S}_j(u)\,du = \sum_{h|t_{jh}<\tau} \left[\hat{S}_j(t_{jh})\left(\min(t_{j,h+1},\tau) - t_{j,h}\right)\right] \quad (3.21)$$

The estimator of the variance of $\hat{\mu}_j$ is given by

$$\hat{V}\left(\hat{\mu}_j\right) = \sum_{i=1}^{n_j} \left(\hat{\xi}_{ji}^{(m)}\right)^2 \quad (3.22)$$

where

$$\hat{\xi}_{ji}^{(m)} = -\left(\frac{I\{X_{ji} \leq \tau\}\delta_{ji}A_j(X_{ji})}{R_{ji}} - \sum_{g=1}^{n_j} \frac{I\{x_{jg} \leq \min(X_{ji},\tau)\}\delta_{jg}A_j(X_{jg})}{R_{jg}^2}\right)$$

$$(3.23)$$

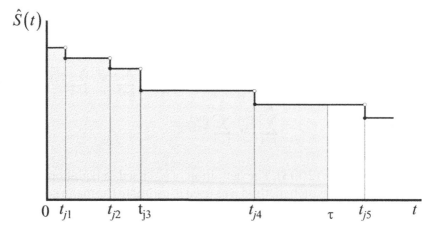

Figure 3.2 Shaded area equal to estimated mean survival time from 0 to τ

and

$$A_j(t) = \int_t^\tau \hat{S}_j(u)du$$

$$= \int_0^\tau \hat{S}_j(u)du - \int_0^t \hat{S}_j(u)du$$

$$= \hat{\mu}_j - \sum_{h|t_{jh}<t} \left[\hat{S}_j(t_{jh})\left(\min(t_{j,h+1}, t) - t_{j,h}\right)\right]$$

see Willan *et al.* (2002). Therefore, when the measure of effectiveness is mean survival time, Δ_e is estimated by

$$\hat{\Delta}_e = \hat{\mu}_T - \hat{\mu}_S \tag{3.24}$$

and the variance of $\hat{\Delta}_e$ is given by

$$\hat{V}\left(\hat{\Delta}_e\right) = \hat{V}\left(\hat{\mu}_T\right) + \hat{V}\left(\hat{\mu}_S\right) = \sum_{i=1}^{n_T}\left(\hat{\xi}_{Ti}^{(m)}\right)^2 + \sum_{i=1}^{n_S}\left(\hat{\xi}_{Si}^{(m)}\right)^2 \tag{3.25}$$

The estimator of the covariance between $\hat{\Delta}_e$ and $\hat{\Delta}_c$ is given by

$$\hat{C}\left(\hat{\Delta}_e, \hat{\Delta}_c\right) = \hat{C}\left(\hat{\mu}_T, \hat{v}_T\right) + \hat{C}\left(\hat{\mu}_S, \hat{v}_S\right) + \sum_{i=1}^{n_T}\left(\hat{\xi}_{Ti}^{(m)}\sum_{k=1}^{K}\hat{\xi}_{Tki}^{(c)}\right)$$

$$+ \sum_{i=1}^{n_S}\left(\hat{\xi}_{Si}^{(m)}\sum_{k=1}^{K}\hat{\xi}_{Ski}^{(c)}\right) \tag{3.26}$$

see Willan *et al.* (2002). The definition of $\hat{\xi}_{jki}^{(c)}$ used in Equation (3.26) will depend on which method was used to estimate the mean cost. The appropriate definition is given in Equation (3.6) for the direct method and in Equation (3.15) for IPW.

3.3.2.2   Inverse-probability weighting

The method of IPW can be used to estimate mean survival time. Let $X_{ji}^* = \min\left(D_{ji}, U_{ji}, \tau\right)$, i.e. X_{ji}^* is the time to death, time to censoring or the duration of interest, whichever is the shortest. Let $\delta_{ji}^* = \delta_{ji} + \bar{\delta}_{ji}I\left\{U_{ji} \geq \tau\right\}$, i.e. δ_{ji}^* equals 1 if, and only if, the patient dies or if the patient is censored after being followed for the duration of interest. The IPW estimator for mean survival time over the duration of interest is given by

$$\hat{\mu}_j = \left(\sum_{i=1}^{n_j} \frac{\delta_{ji}^*}{\hat{G}\left(X_{ji}^*\right)} \right)^{-1} \sum_{i=1}^{n_j} \frac{\delta_{ji}^* X_{ji}^*}{\hat{G}\left(X_{ji}^*\right)}. \tag{3.27}$$

This estimator is simply a weighted average of the times on study i.e. X_{ji}^*) for the patients who are not censored prior to τ. The weight is 1 dividend by the probability of not being censored at the time of death for those who died prior to τ and 1 over the probability of not being censored at τ for those who survived, uncensored, to τ.

The variance of $\hat{\mu}_j$ is given by

$$\hat{V}\left(\hat{\mu}_j\right) = \sum_{i=1}^{n_j} \left(\hat{\xi}_{ji}^{(m)}\right)^2 \tag{3.28}$$

where

$$\hat{\xi}_{ji}^{(m)} = \frac{1}{n_j} \left[\frac{\delta_{ji}^* \left(X_{ji}^* - \hat{\mu}_j\right)}{\hat{G}\left(X_{ji}^*\right)} + \bar{\delta}_{ji} F_{ji} - \sum_{g=1}^{n_j} \frac{\bar{\delta}_{jg} I\left\{X_{jg} \leq X_{ji}\right\}}{R_{jg}} F_{jg} \right] \tag{3.29}$$

and

$$F_{ji} = \frac{1}{R_{ji}} \sum_{g=1}^{n_j} \frac{I\left\{X_{jg}^* > X_{ji}^*\right\} \delta_{jg}^* \left(X_{jg}^* - \hat{\mu}_j\right)}{\hat{G}\left(X_{jg}^*\right)}.$$

The estimator for the covariance between $\hat{\Delta}_e$ and $\hat{\Delta}_c$ is given by

$$\hat{C}\left(\hat{\Delta}_e, \hat{\Delta}_c\right) = \hat{C}\left(\hat{\mu}_T, \hat{v}_T\right) + \hat{C}\left(\hat{\mu}_S, \hat{v}_S\right) = \sum_{i=1}^{n_T} \left(\hat{\xi}_{Ti}^{(m)} \sum_{k=1}^{K} \hat{\xi}_{Tki}^{(c)} \right)$$
$$+ \sum_{i=1}^{n_S} \left(\hat{\xi}_{Si}^{(m)} \sum_{k=1}^{K} \hat{\xi}_{Ski}^{(c)} \right) \tag{3.30}$$

where $\hat{\xi}_{jki}^{(c)}$ is defined in Equation (3.15).

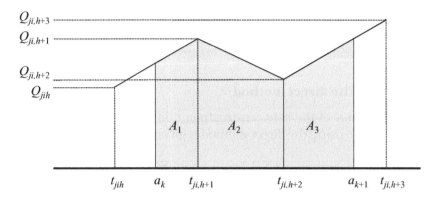

Figure 3.3 QALYs for Kth interval

3.3.3 Mean quality-adjusted survival time

Since patients accrue quality-adjusted survival (QAS) at different rates, much the same as they accrue costs at different rates, the application of life-table methods to estimate mean QAS time produces biased results. So, as with cost, we can utilize either the direct method or inverse-probability weighting to provide estimators for mean QAS time. Recall from Section 2.3.3 that there are m_{ji} QoL measurements on the jith patient, taken at times: $0 \le t_{ji1} < t_{ji2} < \ldots < t_{jim_{ji}} \le \tau$, with corresponding values $Q_{ji1}, Q_{ji2}, \ldots Q_{jim_{ji}}$. The observed QALY on the jith patient during the kth interval, denoted by q_{jki}, is simply the area under the curve of the plot of the Q_{ij} values by the t_{ij} values between a_k and a_{k+1}, as shown in Figure 3.3. Therefore $q_{jki} = \int_{a_k}^{a_{k+1}} Q_{ji}(t)$ dt, where $Q(t)$ is defined as in Equation (2.9.) Put more simply, q_{jik} is the sum of the shaded rectangles shown in Figure 3.3, i.e.

$$q_{jki} = A_1 + A_2 + A_3$$

$$= \left(Q_{ji,h+1} + \frac{\left(Q_{ji,h+1} - Q_{jih} \right) \left(a_k - t_{ji,h+1} \right)}{2 \left(t_{ji,h+1} - t_{jih} \right)} \right) \times \left(t_{ji,h+1} - a_k \right)$$

$$+ \left(Q_{ji,h+1} + Q_{ji,h+2} \right) \left(t_{ji,h+2} - t_{ji,h+1} \right) / 2$$

$$+ \left(Q_{ji,h+2} + \frac{\left(Q_{ji,h+3} - Q_{jih+2} \right) \left(a_{k+1} - t_{ji,h+2} \right)}{2 \left(t_{ji,h+3} - t_{ji,h+2} \right)} \right)$$

$$\times \left(a_{k+1} - t_{ji,h+2} \right)$$

To apply the direct method or IPW, one need only replace C_{jki} by q_{jki} in the appropriate equations in Sections 3.2.1 and 3.2.2, as illustrated in the following sections.

3.3.3.1   The direct method

The estimator of the between-treatment difference of mean quality-adjusted life using the direct method is given by

$$\hat{\Delta}_e = \hat{\varphi}_T - \hat{\varphi}_S = \sum_{k=1}^{K} \hat{S}_T(a_k)\tilde{q}_{Tk} - \sum_{k=1}^{K} \hat{S}_S(a_k)\tilde{q}_{Sk} \qquad (3.31)$$

where $\tilde{q}_{jk} = \left(\sum_{i=1}^{n_j} \delta_{jki}^{**} \right)^{-1} \sum_{i=1}^{n_j} \delta_{jki}^{**} q_{jki}$.

The estimator of the variance of $\hat{\Delta}_e$ is given by, see Willan *et al.* (2003).

$$\hat{V}\left(\hat{\Delta}_e\right) = \hat{V}\left(\hat{\varphi}_T\right) + \hat{V}\left(\hat{\varphi}_S\right) = \sum_{i=1}^{n_T}\sum_{k=1}^{K}\sum_{g=1}^{K} \hat{\xi}_{Tki}^{(q)}\hat{\xi}_{Tgi}^{(q)} + \sum_{i=1}^{n_S}\sum_{k=1}^{K}\sum_{g=1}^{K} \hat{\xi}_{Ski}^{(q)}\hat{\xi}_{Sgi}^{(q)}$$

$$(3.32)$$

where

$$\hat{\xi}_{jki}^{(q)} = \left(\sum_{i=1}^{n_j} \delta_{jki}^{**} \right)^{-1} \left(q_{ijk} - \tilde{q}_{jk} \right) \hat{S}_j\left(a_k\right)\delta_{jki}^{**}$$

$$-\hat{S}_j\left(a_k\right)\tilde{q}_{jk} \left(\frac{I\left\{X_{ji} \le a_k\right\}\delta_{ji}}{R_{ji}} - \sum_{g=1}^{n_j} \frac{I\left\{X_{jg} \le \min\left(X_{ji}, a_k\right)\delta_{jg}\right\}}{R_{jg}^2} \right)$$

$$(3.33)$$

The estimator for the covariance between $\hat{\Delta}_e$ and $\hat{\Delta}_c$ is given by

$$\hat{C}\left(\hat{\Delta}_e, \hat{\Delta}_c\right) = \hat{C}\left(\hat{\varphi}_T, \hat{v}_T\right) + \hat{C}\left(\hat{\varphi}_S, \hat{v}_S\right) = \sum_{i=1}^{n_T}\sum_{k=1}^{K}\sum_{g=1}^{K} \hat{\xi}_{Tki}^{(q)}\hat{\xi}_{Tgi}^{(c)}$$

$$+ \sum_{i=1}^{n_S}\sum_{k=1}^{K}\sum_{g=1}^{K} \hat{\xi}_{Ski}^{(q)}\hat{\xi}_{Sgi}^{(c)} \qquad (3.34)$$

where $\hat{\xi}_{jki}^{(c)}$ is defined in Equation (3.6.) The reader is referred to Willan *et al.* (2003) for more details.

3.3.3.2 Inverse-probability weighting

The estimator of the between-treatment difference of mean quality-adjusted life using IPW is given by

$$\hat{\Delta}_e = \hat{\varphi}_T - \hat{\varphi}_S = \sum_{k=1}^{K} \hat{\varphi}_{Tk} - \sum_{k=1}^{K} \hat{\varphi}_{Sk} \qquad (3.35)$$

where

$$\hat{\varphi}_{jk} = \left(\sum_{i=1}^{n_j} \frac{\delta_{jki}^*}{G\left(X_{jki}^*\right)} \right)^{-1} \sum_{i=1}^{n_j} \frac{\delta_{jki}^* q_{jki}}{G\left(X_{jki}^*\right)} \qquad (3.36)$$

The variance of $\hat{\Delta}_e$ is given by

$$\hat{V}\left(\hat{\Delta}_e\right) = \hat{V}\left(\hat{\varphi}_T\right) + \hat{V}\left(\hat{\varphi}_S\right) = \sum_{i=1}^{n_T} \sum_{k=1}^{K} \sum_{g=1}^{K} \hat{\xi}_{Tki}^{(q)} \hat{\xi}_{Tgi}^{(q)} + \sum_{i=1}^{n_S} \sum_{k=1}^{K} \sum_{g=1}^{K} \hat{\xi}_{Ski}^{(q)} \hat{\xi}_{Sgi}^{(q)}$$

$$(3.37)$$

where

$$\hat{\xi}_{jki}^{(q)} = \frac{1}{n_j} \left(\frac{\delta_{jki}^* \left(q_{jki} - \hat{\varphi}_{jk} \right)}{\hat{G}\left(X_{jki}^*\right)} + \bar{\delta}_{ji} D_{jki} - \sum_{g=1}^{n_j} \frac{\bar{\delta}_{jg} I\left\{X_{jg} \le X_{ji}\right\} D_{jkg}}{R_{jg}} \right)$$

$$(3.38)$$

and $D_{jki} = \dfrac{1}{R_{ji}} \displaystyle\sum_{g=1}^{n_j} \dfrac{I\left\{X_{jkg}^* > X_{ji}\right\} \delta_{jkg}^* \left(q_{jkg} - \hat{\varphi}_{jk}\right)}{\hat{G}\left(X_{jkg}^*\right)}$. Again, to ease

the burden of notation, $\hat{\xi}_{jki}^{(q)}$ has two definitions, one in Equation (3.32) for the direct method and one in Equation (3.37), for the IPW method. The estimator for the covariance between $\hat{\Delta}_e$ and $\hat{\Delta}_c$ is given by

$$\hat{C}\left(\hat{\Delta}_e, \hat{\Delta}_c\right) = \hat{C}\left(\hat{\varphi}_T, \hat{v}_T\right) + \hat{C}\left(\hat{\varphi}_S, \hat{v}_S\right) = \sum_{i=1}^{n_T} \sum_{k=1}^{K} \sum_{g=1}^{K} \hat{\xi}_{Tki}^{(q)} \hat{\xi}_{Tgi}^{(c)}$$

$$+ \sum_{i=1}^{n_S} \sum_{k=1}^{K} \sum_{g=1}^{K} \hat{\xi}_{Ski}^{(q)} \hat{\xi}_{Sgi}^{(c)} \qquad (3.39)$$

where $\hat{\xi}_{jki}^{(c)}$ is defined in Equation (3.15), see Willan *et al.* (2002).

3.4 SUMMARY

Methods for estimating the five parameters required for a cost-effectiveness comparison of two groups using censored data are given in Sections 3.2 and 3.3. Life-table methods are used for the probability of surviving and the mean survival time. Inverse-probability weighting is used for mean cost, mean survival time and mean quality-adjusted survival time. The direct (Lin) method is used for mean cost and mean quality-adjusted survival time. In general, inverse-probability weighting is preferred to the direct method because it provides consistent estimators regardless of the censoring pattern. Furthermore, inverse-probability weighting can be generalized to include covariate adjustment, as discussed in Chapter 7. A discussion on the use of these parameters in cost-effectiveness analysis is provided in the following chapter. The use of incremental cost-effectiveness ratios, incremental net benefit and Bayesian methods are discussed.

4

Cost-effectiveness Analysis

4.1 INTRODUCTION

Statistical inference in cost-effectiveness analysis, using the five parameter estimates calculated from the methods given in Chapters 2 and 3, are presented in this chapter. Traditionally, cost-effectiveness analysis has summarized the value for money of the treatment under evaluation using the incremental cost-effectiveness ratio (ICER). Statistical inference using the ICER is covered in Section 4.2, focusing on the many difficulties and complexities that arise when using ratio statistics. The concerns regarding inference on the ICER have led investigators to the use incremental net benefit (INB) as an alternative. Statistical inference based on INB is discussed in Section 4.3. The equivalence of the appropriate inference using either statistic is emphasized through the equivalence of the regions on the cost-effectiveness plane. The presentation of results using the cost-effectiveness acceptability curve (CEAC) is presented in Section 4.4 and the use of bootstrap methods is discussed in Section 4.5. The Bayesian approach, under which the CEAC has its most natural interpretation, is presented in Section 4.6. Lastly, in Section 4.7 consideration is given to the fact that the decision threshold may differ, depending on whether a policy decision involves implementing a new treatment or withdrawing an existing treatment.

Statistical Analysis of Cost-effectiveness Data. A. Willan and A. Briggs
© 2006 John Wiley & Sons, Ltd.

4.2 INCREMENTAL COST-EFFECTIVENESS RATIO

The ICER is estimated by $\hat{R} = \hat{\Delta}_c / \hat{\Delta}_e$ = slope of $\hat{\Delta}$, where $\hat{\Delta} = (\hat{\Delta}_e, \hat{\Delta}_c)^T$. The estimator $\hat{R}$ is consistent, although not unbiased. In a standard test of hypothesis approach one looks for evidence to reject the following null hypothesis:

$$\text{H}: \quad \frac{\Delta_c}{\Delta_e} \geq \lambda \quad \text{if } \Delta_e > 0; \quad \text{or} \quad \frac{\Delta_c}{\Delta_e} \leq \lambda \quad \text{if } \Delta_e < 0 \quad (4.1)$$

in favor of

$$\text{A}: \quad \frac{\Delta_c}{\Delta_e} < \lambda \quad \text{if } \Delta_e > 0; \quad \text{or} \quad \frac{\Delta_c}{\Delta_e} > \lambda \quad \text{if } \Delta_e < 0. \quad (4.2)$$

These rather complex hypotheses result from the ICER being a ratio and this is one of the arguments in favor of using an analysis based on INB (see Section 4.3). Nevertheless, the reason for this formulation of the cost-effectiveness hypothesis is clear from the cost effectiveness plane (introduced in Figure 1.1). Figure 4.1 shows the cost-effectiveness plane including the threshold line with slope λ and passing through the origin. The shaded area below and to the right of this line is the acceptance region (corresponding to the alternative hypothesis of Equation 4.2) and the unshaded area above and to the left of the threshold line is the rejection region (corresponding to the null hypothesis of Equation 4.1). The corresponding ICERs for points a_1 and a_2 are both less than the threshold, but a_1 falls in the rejection region and a_2 in the acceptance region. Similarly, the corresponding ICERs for b_1 an b_2 are both greater than the threshold, but with opposing decision implications. In order to understand this apparent anomaly it is necessary to consider the denominator of the ratio. Where Δ_e is positive (the NE and SE quadrants in Figure 4.1) Treatment is cost-effective if, and only if, the ICER $< \lambda$. However, where Δ_e is negative (the NW and SW quadrants) Treatment is cost-effective if, and only if the ICER $> \lambda$.

Inference for the ICER is limited to constructing its confidence interval $\left(R_{2\alpha}^{LL}, R_{2\alpha}^{UL} \right)$ where the superscript refers to the lower and upper

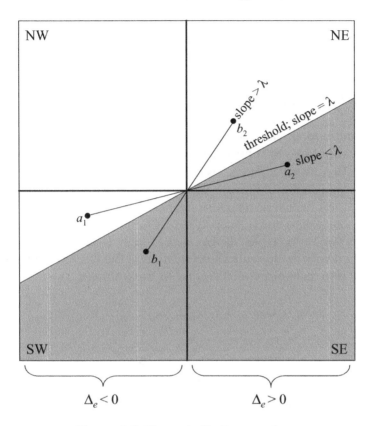

Figure 4.1 The cost-effectiveness plane

limit respectively and 2α is the level that provides a $(1 - 2\alpha)100\%$ confidence interval. The conclusion that there is evidence that Treatment is cost-effective is made if and only if either: (i) $\hat{\Delta}_e > 0$ and $R_{2\alpha}^{UL} < \lambda$; or (ii) $\hat{\Delta}_e < 0$ and $R_{2\alpha}^{LL} > \lambda$.

However, confidence interval construction for the ICER is not straightforward. In particular, Δ_e can be 0, which means that R is undefined and therefore not properly estimable. Some authors have proposed using a Taylor series expansion to estimate the variance of $\hat{R}$, see O'Brien *et al.* (1994), Chaudhary and Stearns (1996), Briggs and Fenn (1998) and Briggs *et al.* (1999). Using the Taylor series expansion the estimator for the variance of $\hat{R}$ is given by

$$\hat{V}\left(\hat{R}\right) \simeq \hat{R}^2 \left[\hat{V}\left(\hat{\Delta}_e\right)/\hat{\Delta}_e^2 + \hat{V}\left(\hat{\Delta}_c\right)/\hat{\Delta}_c^2 - 2\hat{C}\left(\hat{\Delta}_e, \hat{\Delta}_c\right)/\hat{\Delta}_e\hat{\Delta}_c\right]$$

and $100\left(1 - 2\alpha\right)\%$ confidence interval can be constructed by $\hat{R} \pm z_{1-\alpha}\sqrt{\hat{V}(\hat{R})}$, where $z_{1-\alpha}$ is the $100(1 - \alpha)$th percentile of a standard normal random variable. However, the accuracy of this approximation cannot be relied on when the sample sizes are small, or when $\hat{\Delta}_e/\sqrt{\hat{V}(\hat{\Delta}_e)}$ or $\hat{\Delta}_c/\sqrt{\hat{V}(\hat{\Delta}_c)}$ is less than 0.1.

An application of Fieller's theorem provides an alternative to the use of the Taylor series expansion for calculating confidence intervals for R, see Willan and O'Brien (1996), Chaudhary and Sterns (1996), Briggs and Fenn (1998), Briggs *et al.* (1999), Laska EM *et al.* (1997). The best way to illustrate the derivation of the Fieller confidence interval is to recognize that if $\hat{\Delta}_e$ and $\hat{\Delta}_c$ are unbiased, then

$$E(\hat{\Delta}_c - R\hat{\Delta}_e) = E(\hat{\Delta}_c) - \frac{\Delta_c}{\Delta_e}E(\hat{\Delta}_e)$$

$$= \Delta_c - \frac{\Delta_c}{\Delta_e}\Delta_e$$

$$= 0$$

and, if $\hat{\Delta}_e$ and $\hat{\Delta}_c$ are normally distributed, then so is $\hat{\Delta}_c - R\hat{\Delta}_e$, and therefore

$$\left(\hat{\Delta}_c - R\hat{\Delta}_e\right)\bigg/\sqrt{V\left(\hat{\Delta}_c - R\hat{\Delta}_e\right)}$$

$$= \left(\hat{\Delta}_c - R\hat{\Delta}_e\right)\bigg/\sqrt{V\left(\hat{\Delta}_c\right) + R^2 V\left(\hat{\Delta}_e\right) - 2R \times C\left(\hat{\Delta}_c, \hat{\Delta}_e\right)}$$

is distributed as a standard normal random variable. Consequently,

$$\left|\hat{\Delta}_c - R\hat{\Delta}_e\right|\bigg/\sqrt{V\left(\hat{\Delta}_c\right) + R^2 V\left(\hat{\Delta}_e\right) - 2R \times C\left(\hat{\Delta}_c, \hat{\Delta}_e\right)} \leq z_{1-\alpha}$$

$$(4.3)$$

with probability $(1 - 2\alpha)$. By substituting in the estimates for the variances and the covariance and squaring both sides, Equation (4.3) becomes a quadratic equation in R. Solving for R provides the $100(1 - 2\alpha)\%$ confidence interval for ICER, the limits of which are given by

$$\hat{R}\left\{\left(1 - z_{1-\alpha}^2 c \pm z_{1-\alpha}\sqrt{a + b - 2c - z_{1-\alpha}^2(ab - c^2)}\right)\Big/\left(1 - z_{1-\alpha}^2 a\right)\right\}$$

(4.4)

where $a = \hat{V}(\hat{\Delta}_e)/\hat{\Delta}_e^2$, $b = \hat{V}(\hat{\Delta}_c)/\hat{\Delta}_c^2$ and $c = \hat{C}(\hat{\Delta}_e, \hat{\Delta}_c)/(\hat{\Delta}_e \hat{\Delta}_c)$.

The Fieller confidence interval is a 'bow tie' defined by the two lines through the origin and includes the wedge in which the observed point lies and the wedge opposite, see Figure 4.2 for illustration. This is

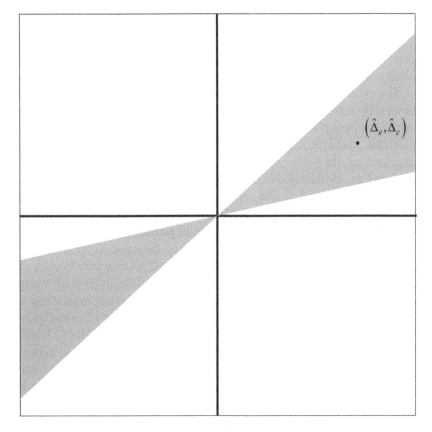

Figure 4.2 Bow tie ICER confidence region

because the slope of any point in either wedge satisfies Equation (4.3). This is particularly vexing when the observed point is in one of the non-trade-off quadrants (i.e. S.E. or N.W.) because then the confidence interval will include parts from both the win–win and lose–lose regions of the cost-effectiveness plane. In practice, when the observed point is sufficiently far away from the origin, and the confidence interval is narrow, very little probability will lie in the opposite wedge.

The confidence interval can include part of the vertical axis, in which case the corresponding limit is considered to be arbitrarily large if it is the upper limit and arbitrarily small if it is the lower limit. If $a + b - 2c - z_{1-\alpha}^2(ab - c^2) = 0$, the upper and lower limits are the same and are equal to $\hat{R}(1 - z_{1-\alpha}^2 c)/(1 - z_{1-\alpha}^2 a)$, in which case, the confidence interval defines the entire CE plane. If $a + b - 2c - z_{1-\alpha}^2(ab - c^2) < 0$, no solution exists, and again the confidence interval defines the entire CE plane. Such are the problems when dealing with ratio statistics.

The extent to which the Taylor series expansion is an approximation is seen by rewriting it as $\hat{R}(1 \pm z_{1-\alpha}\sqrt{a + b - 2c})$ and contrasting it with Expression (4.4). The divisor $(1 - z_{1-\alpha}^2 a)$ tends to dominate the contrast, and although it is not obvious from routine inspection of the expressions, the Fieller solution will always provide a wider confidence interval, see Willan and O'Brien (1996). In general the Fieller solution is preferred since it is not an approximation and relies only on the estimators $\hat{\Delta}_e$ and $\hat{\Delta}_c$ being unbiased and normally distributed. As was discussed in Chapter 2, cost data in particular may often exhibit substantial skewness such that there may be some concern over the legitimacy of the normal assumption for $\hat{\Delta}_c$. However, Willan and O'Brien (1996) provide simulation evidence that, even with sample sizes as small as 100 and cost data that are log-normal to the base 10, the distribution of $\hat{\Delta}_c$ is normally distributed. Other studies have come to the same conclusion, see Thompson and Barber (2000), Lumley *et al.* (2002), Willan *et al.* (2004), and Briggs *et al.* 2005. In general, these simulation studies support Cochran's rule of thumb that n must be $25\eta^2$ for situations where the 'principal deviation from normality consists of marked positive skewness' (Cochran 1977) and where η is the skewness coefficient in the sample. The guideline was devised such that a 95% confidence interval will have an error probability no greater than 6%.

Table 4.1   Three treatment comparison falling in the win–win quadrant

	Δ_e	Δ_c	ICER
T1	0.1	-2000	$-20\,000$
T2	0.2	-4000	$-20\,000$
T3	0.2	-2000	$-10\,000$

The ICER is not properly ordered in the SE and NW quadrants, i.e. in the win–win and lose–lose quadrants, respectively. To see this, consider comparing S to three different treatments ($T1$, $T2$ and $T3$) for which the values of Δ_e and Δ_c are shown in Table 4.1. The measure of effectiveness is the probability of survival. $T1$ increases the probability of survival by 0.1 and reduces the cost by 2000, for an ICER of $-20\,000$. $T2$ increases the probability of survival by 0.2 and reduces the cost by 4000, and is clearly superior to $T1$, yet it has the same ICER. Therefore, two treatments compared to the same standard can have the same ICER, yet one may be clearly superior to the other. It gets worse. $T3$ increases the probability of survival by 0.2 and reduces the cost by 2000, and is superior to $T1$ because, although it reduces cost by the same amount, its increase in the probability of survival is double. However, the ICER for $T3$ is larger than for $T1$. It is not possible to interpret the confidence interval for a parameter over a range in which it is not properly ordered, and this concern is one more argument in favour of an INB approach.

4.3   INCREMENTAL NET BENEFIT

The concerns encountered when making statistical inference on the ICER can be addressed by using an INB approach. The estimator of INB is given by $\hat{b}_\lambda = \hat{\Delta}_e \lambda - \hat{\Delta}_c$, and is unbiased if $\hat{\Delta}_e$ and $\hat{\Delta}_c$ are. The variance of $\hat{b}_\lambda$ is estimated by $v_\lambda \equiv \hat{V}(\hat{b}_\lambda) = \lambda^2 \hat{V}(\hat{\Delta}_e) + \hat{V}(\hat{\Delta}_c) - 2\lambda\hat{C}(\hat{\Delta}_e, \hat{\Delta}_c)$. The null hypothesis H: $\Delta_e\lambda - \Delta_c \leq 0$ can be rejected in favour of the alternative hypothesis A: $\Delta_e\lambda - \Delta_c > 0$ at the level α if the test statistic $\hat{b}_\lambda/\sqrt{v_\lambda}$ exceeds $z_{1-\alpha}$. Additionally, the $100(1 - 2\alpha)\%$

confidence limits for INB are given by

$$\hat{b}_\lambda \pm z_{1-\alpha}\sqrt{\hat{v}_\lambda} \qquad\qquad (4.6)$$

and H can be rejected in favor of A, equivalently, if the lower limit exceeds 0. Note that the hypotheses H and A are, respectively, exactly the same as the hypotheses defined in Equations (4.1) and (4.2). In terms of Figure 4.1, note that the acceptance and rejection regions relating to the INB formulation exactly correspond to those under the ICER interpretation with points above and to the left of the threshold line associated with INB < 0 (i.e. not cost-effective) and those below and to the right having INB > 0 (i.e. cost-effective). It should be straightforward to appreciate the ease with which the INB approach can test H in comparison with the ICER approach.

Another advantage to the INB approach is that a sensitivity analysis, varying the value of λ, can be performed. Therefore, by estimating $\hat{b}_\lambda$ and calculating the corresponding confidence limits for a large number of values of λ, and plotting them as a function of λ, as seen in Figure 4.3, readers can apply the value of λ they feel most appropriate, reading from the vertical axis the estimate of INB and the confidence limits. Furthermore, $\hat{b}_\lambda$ crosses the horizontal axis at $\hat{R}$ (i.e. $\hat{b}_{\hat{R}} = 0$), and the confidence limits for b_λ crosses the horizontal axis at the Fieller limit for R, allowing readers to make inference on the ICER. This can be seen by setting Expression (4.6) to 0, which becomes the same equation in λ as the Inequality (4.3) is in R. The correspondence between the Fieller and net-benefit solution has been noted in the literature (Heitjan, 2000; Zethraeus and Lothgren, 2000), but should be unsurprising given the similar configurations of $\hat{\Delta}_e$ and $\hat{\Delta}_c$ and the assumption of joint-normality employed in both approaches. In Figure 4.3 the lower limit for b_λ fails to cross the horizontal axis then the upper limit for the ICER does not exist and is considered to be arbitrarily large. If the upper limit for b_λ fails to cross the horizontal axis then the lower limit for the ICER does not exist and is considered to be arbitrarily small. If neither limit for b_λ crosses the horizontal axis then neither limit for the ICER exists because, as discussed in Section 4.2, $a + b - 2c - z_{1-\alpha}^2(ab - c^2) = 0$. Finally, $\hat{b}_\lambda$ crosses the vertical axis at $-\hat{\Delta}_c$, and the lower and upper limits for b_λ cross the vertical axis at the lower and upper limit for $-\hat{\Delta}_c$ allowing inference on Δ_c.

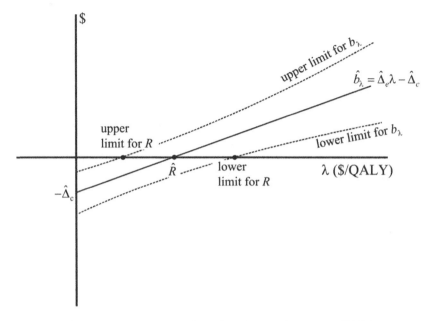

Figure 4.3 Incremental net benefit as a function of WTP

It should also be straightforward to see that the INB could be boot-strapped as an alternative to assuming a normal distribution. While the necessity for bootstrapping the INB, given that it does not suffer from the problems associated with ratio statistics is less clear, it is worth noting that a bootstrap analysis of net benefit and an appropriate bootstrap analysis of the ICER should exactly correspond in terms of inference regarding cost-effectiveness, just as the parametric INB and parametric Fieller method correspond.

4.4 THE COST-EFFECTIVENESS ACCEPTABILITY CURVE

For the methods of inference discussed in Sections 4.2 and 4.3 the level of significance is fixed at some level α. This may not always be appropriate, and a Bayesian approach leading to the cost-effectiveness acceptability curve (CEAC) can provide a more flexible and a natural means for expressing the uncertainty of parameter estimators, see van Hout *et al.* (1994), Briggs and Fenn (1998) and Briggs (1999). The CEAC is a plot of the probability that Treatment is cost-effective as a function of willingness-to-pay. In Bayesian terms, it is the probability that the INB, for a particular value of the willingness-to-pay, is greater than zero, (i.e. $\Pr(\Delta_e \lambda - \Delta_c > 0)$, and is the probability that the point Δ falls below the line on the CE plane that passes through the origin with slope λ (the threshold), see Figure 4.4 for illustration. The concentric ellipses represent contours of equal probability for the

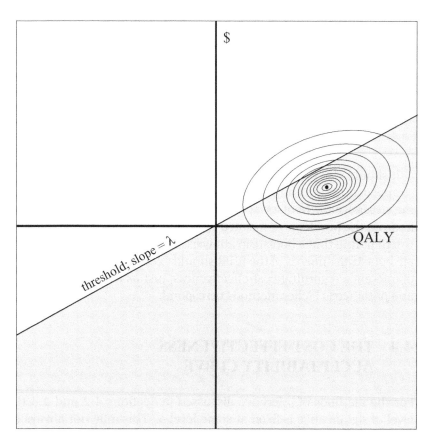

Figure 4.4 The CE plane; shaded area $= \Pr(b_\lambda > 0)$

joint posterior density function of Δ. The CEAC is the probability that the point Δ lies below the threshold as a function of λ. The CEAC = $\Pr(\Delta_c < 0)$ for, $\lambda = 0$ approaches $\Pr(\Delta_e < 0)$ as λ approaches $-\infty$ and approaches $\Pr(\Delta_e > 0)$ as λ approaches $+\infty$. The probabilities are determined from the posterior distribution of Δ, which is, if an uninformative prior is specified, the distribution of $\hat{\Delta}$. If normality is assumed, then the posterior distribution of Δ will be bivariate normal with mean $\hat{\Delta}$ variance–covariance matrix

$$\hat{V}(\hat{\Delta}) = \begin{pmatrix} \hat{V}(\hat{\Delta}_e) & \hat{C}(\hat{\Delta}_e, \hat{\Delta}_c) \\ \hat{C}(\hat{\Delta}_e, \hat{\Delta}_c) & \hat{V}(\hat{\Delta}_c) \end{pmatrix}$$

If we let $f_{\hat{\Delta}}(\cdot, \cdot)$ be the corresponding density function, then the CEAC for a particular value of λ is given by

$$\mathcal{A}(\lambda) \equiv \int_{-\infty}^{\infty} \int_{-\infty}^{\lambda e} f_{\hat{\Delta}}(e, c)\, dc\, de \;=\; \Phi(\hat{b}(\lambda)/\sqrt{v_\lambda}) \qquad (4.7)$$

where $\Phi(\cdot)$ is the cumulative distribution function for a standard normal random variable. $\mathcal{A}(\lambda)$ has a frequentist interpretation as 1 minus the p-value for the test of hypothesis H: $\Delta_e \lambda - \Delta_c \leq 0$ versus A: $\Delta_e \lambda - \Delta_c > 0$. Clearly, $\mathcal{A}(\hat{R}) = 0.5$. In addition, the CEAC passes through Fieller limits for the ICER at α and $1 - \alpha$. If the CEAC does not pass through α or $1 - \alpha$, then the corresponding limit for the ICER does not exist and is considered to be arbitrarily large or small because, as described in Section 4.2, the interval includes the vertical axis. If the CEAC passes through neither α nor $1 - \alpha$, then neither limit for the ICER exists because, as discussed in Section 4.2, $a + b - 2c - z_{1-\alpha}^2(ab - c^2) < 0$. As before, the confidence limits for the ICER derived from the CEAC define a 'bow tie' region as illustrated in Figure 4.2.

The CEAC has three important strengths. It is a measure of both magnitude and uncertainty of cost-effectiveness. It expresses the uncertainty in terms of a probability statement about the cost-effectiveness of Treatment, which is often considered more natural to policymakers. Thirdly, it allows policymakers the ability to use different strengths of evidence depending on what Treatment is. If adopting Treatment is expected to have a substantial effect on the health care

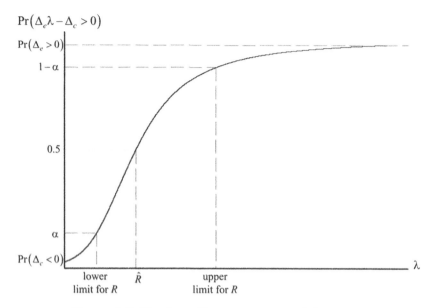

Figure 4.5 The Cost-effectiveness acceptability curve

system, because either Δ_c or the population to which it applies is large, then policymakers might want less uncertainty, i.e. a high value for $\mathcal{A}(\lambda)$, perhaps 0.99. Conversely, if Treatment is expected to have very little impact, lower values of $\mathcal{A}(\lambda)$ might be tolerated, perhaps 0.8.

Although the CEAC in Figure 4.5 looks similar to a probability distribution function, in practice a CEAC can take on a variety of shapes, including those with negative slope. For a more complete discussion on the CEAC and the various shapes that it can take, see Fenwick *et al.* (2004).

4.5 USING BOOTSTRAP METHODS

Bootstrapping has been proposed as another approach for calculating the CEAC and confidence limits for the ICER, see Chaudhary and Sterns (1996), Briggs *et al.* (1997), Briggs and Fenn (1998), Briggs *et al.* (1999). For the bootstrap approach the analyst re-samples the data at random with replacement, sampling the same number of observations as are in the original data set, i.e. n_T from arm T and n_S from arm S. Consequently, some patients are sampled more than once and some patients not at all. The procedure is repeated so that the data are re-sampled B times. The validity of the bootstrap approach rests on two asymptotics: (i) as the original sample size approaches the population size so the sample distribution tends to the population distribution; and, given this, (ii) as B, the number of bootstrap replications approaches infinity so the bootstrap estimate of the sampling distribution of a statistic approaches the true sampling distribution (Mooney and Duval, 1993). The analyst can then calculate the estimates for Δ_e, Δ_c and the ICER for the re-sampled data; these are denoted as $\hat{\Delta}_{ei}^*$, $\hat{\Delta}_{ci}^*$ and R_i^*, $i = 1, 2, \ldots B$, respectively. The set of values of R_i^* provide an estimator for the distribution of $\hat{R}$, the estimator of the ICER from the original set of data.

Efron and Tibshirani (1993) suggest that the 'ideal' bootstrap estimate corresponds to an infinite number of bootstrap re-samples. In practice, although there are no formal rules regarding the number of bootstrap replications required for reliable estimation, they suggest that 50 replications are usually adequate to provide an informative estimate of variance and that very seldom are more than 200 replications required. For the estimation of percentile confidence intervals, more replications are required in order to better estimate the tails of the sampling distribution. It is generally agreed that between 2000 and 5000 re-samples are sufficient to achieve stability for percentile interval estimates. A number of different methods of confidence interval estimation based on bootstrap replication are possible including: normal approximation, percentile, bias-corrected and accelerated percentile and percentile-t (Efron and Tibshirani 1993; Mooney and Duval 1993). However, the normal approximation and percentile-t methods make use of the bootstrap estimate of variance of the ICER which is not recommended because, as ratios, the values of R_i^* are not normally distributed, and extreme values occur when $\hat{\Delta}_{ei}^*$ approaches zero. The bias corrected and accelerated methods adjust for the inherent bias and skewness of ratio statistics. Nevertheless, we concentrate here on the straightforward percentile method, since, as we will go on to argue, this method is consistent with decision making on the

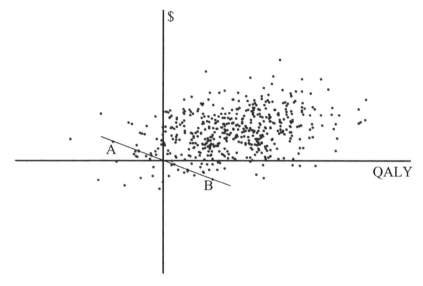

Figure 4.6 Plot of re-sampled $(\hat{\Delta}^*_{ei}, \hat{\Delta}^*_{ci})$

cost-effectiveness plane and provides equivalent results using INB and acceptability curve approaches to handling uncertainty. For more details regarding general bootstrap methods the reader is referred to Efron and Tibshirani (1993). For specific details on the bootstrap estimation of confidence intervals for ICERs, including comparison of the four approaches to interval estimation outlined above, the reader is referred to Briggs *et al.* (1997) and Briggs *et al.* (1999).

An illustration of how the statistics from the re-samples are used to define a percentile confidence interval for R is given in Figure 4.6. Each point on the CE plane in Figure 4.6 represents a $(\hat{\Delta}^*_{ei}, \hat{\Delta}^*_{ci})$ pair from one of the re-samples. To construct the $100(1 - 2\alpha)\%$ confidence interval the analyst must find two rays from the origin that enclose $\hat{\Delta}$ from the original data and $100(1 - 2\alpha)\%$ of the $(\hat{\Delta}^*_{ei}, \hat{\Delta}^*_{ci})^T$ points from the re-samples. The percentile method is based on an ordering of the $(\hat{\Delta}^*_{ei}, \hat{\Delta}^*_{ci})^T$ points. The ordering cannot be based on the R^*_i values because two different re-samplings of the data can have the same R^*_i, but be from different quadrants, and in the case of negative values can have totally different interpretations, one being in the win–win SE quadrant and the other in the lose–lose NW quadrant. For example, consider points A and B in Figure 4.6. They have the same slope, but one (A) is a candidate to be in the upper tail and the other (B) in the lower. Whenever the re-samples lie on both sides of the vertical axis of the CE plane, confidence limits based on orderings of R_i^* will have inappropriate coverage. This point has often been overlooked in theoretical derivations and applications of bootstrap in cost-effectiveness analysis.

As an alternative to deriving bootstrap confidence limits for the ICER we first derive the CEAC from the bootstrap re-samples; so that

$$\mathcal{A}^{\mathrm{B}}(\lambda) \equiv \sum_{i=1}^{B} I\left\{ \hat{\Delta}_{ei}^* \lambda - \hat{\Delta}_{ci}^* > 0 \right\} \bigg/ B$$

and the confidence limits are given by $\mathcal{A}^{\mathrm{B}}(\alpha)$ and $\mathcal{A}^{\mathrm{B}}(1 - \alpha)$. As before, these limits define a 'bow tie' region as illustrated in Figure 4.2.

4.6   A BAYESIAN INCREMENTAL NET BENEFIT APPROACH

A Bayesian approach for INB can be employed by determining its posterior distribution, which is defined as $\mathcal{P}_\lambda(b) \equiv \mathrm{Pr}(b_\lambda \leq b)$. For an uninformative prior, and assuming normality, the posterior distribution of INB is normal with mean $\hat{b}_\lambda$ and variance v_λ, and if we let $f_{b_\lambda}(\cdot)$ be the corresponding density function, then

$$\mathcal{P}_\lambda(b) = \int_{-\infty}^{b} f_{b_\lambda}(x)\,\mathrm{d}x = \Phi\left((b - \hat{b}_\lambda)/\sqrt{v_\lambda} \right) \qquad (4.8)$$

where $\Phi(\cdot)$ is the cumulative distribution function for a standard normal random variable. $\mathcal{P}_\lambda(b)$ is also the probability that the point $\hat{\Delta}$ lies above the line on the CE plane with slope λ and vertical intercept $-b$, since for all points above this line INB $< b$, see Figure 4.7. Again, the concentric ellipses represent contours of equal probability for the density function $f_{\hat{\Delta}}(\cdot, \cdot)$, and an alternative expression for $\mathcal{P}_\lambda(b)$ is

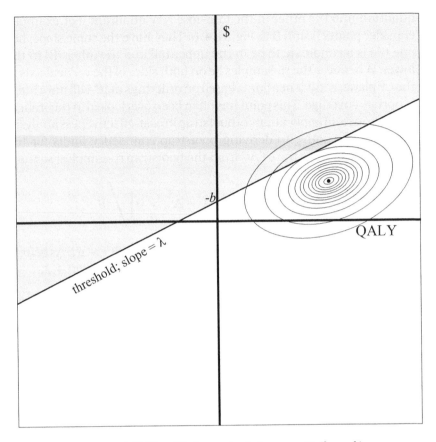

Figure 4.7 The CE plane; shaded area $= \Pr(b_\lambda < b)$

given by

$$\mathcal{P}_\lambda(b) = \int\limits_{-\infty}^{\infty} \int\limits_{e\lambda-b}^{\infty} f_{\hat{\Delta}}(e, c) \, \mathrm{d}c \, \mathrm{d}e \qquad (4.9)$$

Alternatively, bootstrapping can be employed; so that

$$\mathcal{P}_\lambda(b) = \frac{\sum\limits_{i=1}^{B} I\left\{ \hat{\Delta}_{ei}^* \lambda - \hat{\Delta}_{ci}^* \leq b \right\}}{B}$$

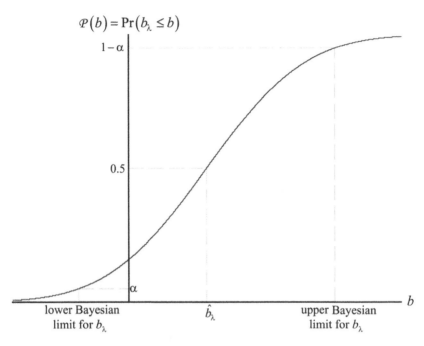

Figure 4.8 The posterior distribution for b_λ

The Bayesian estimate of INB is that value of b for which $\mathcal{P}_\lambda(b) = 0.5$, and the Bayesian lower and upper limits are the values of b for which $\mathcal{P}_\lambda(b) = \alpha$ and $\mathcal{P}_\lambda(b) = 1 - \alpha$, respectively. With an uninformative prior, $\mathcal{P}_\lambda(\hat{b}_\lambda) = 0.5$, $\mathcal{P}_\lambda(\hat{b}_\lambda - z_{1-\alpha}\sqrt{v_\lambda}) = \alpha$ and $\mathcal{P}_\lambda(\hat{b}_\lambda + z_{1-\alpha}\sqrt{v_\lambda}) = 1 - \alpha$, and the Bayesian and frequentist solution are the same, see Figure 4.8.

If the analyst has information on which to base a prior distribution for b_λ, a full Bayesian analysis is possible. If the prior distribution is normal with mean $b_{0\lambda}$ and variance $v_{0\lambda}$, then the posterior distribution for b_λ, given the data, is normal with mean $b_{1\lambda}$ and variance $v_{1\lambda}$, where

$$b_{1\lambda} = v_{1\lambda}\left(\frac{b_{0\lambda}}{v_{0\lambda}} + \frac{\hat{b}_\lambda}{\hat{v}_\lambda}\right) \qquad \text{and } v_{1\lambda} = \left(\frac{1}{v_{0\lambda}} + \frac{1}{\hat{v}_\lambda}\right)^{-1}$$

and if $f_{1b_\lambda}(\cdot)$ is the corresponding density function, then $\mathcal{P}_\lambda(b) = \int_{-\infty}^{b} f_{1b_\lambda}(x)\,dx = \Phi((b - b_{1\lambda})/\sqrt{v_{1\lambda}})$.

The estimator of b_λ and the lower and upper Bayesian limits are the values of b that satisfy $\mathcal{P}_\lambda(b) = 0.5$, $\mathcal{P}_\lambda(b) = \alpha$ and $\mathcal{P}_\lambda(b) = 1 - \alpha$, respectively, and are given by $b_{1\lambda}$, $b_{1\lambda} - z_{1-\alpha}\sqrt{v_{1\lambda}}$ and $b_{1\lambda} + z_{1-\alpha}\sqrt{v_{1\lambda}}$, respectively. Performing these calculations for numerous values of λ in a sensitivity analysis will yield graphs as shown in Figure 4.3. The horizontal intercepts of these graphs provide posterior estimates and Bayesian limits for R, while the negative intercepts of the vertical axis provide posterior estimates and Bayesian limits for Δ_c.

4.7   KINKED THRESHOLDS

Under standard principles and assumptions of microeconomic theory, there should be little difference between a person's WTP for a commodity or program and the compensation they would demand—willingness to accept (WTA)—to relinquish the same commodity, see Johansson (1995). In practice there has been a wide and reproducible disparity between measured WTP and WTA values. In a recent meta-analysis of published studies by O'Brien *et al.* (2002), the ratio of WTA to WTP is approximately 7 for environmental studies and approximately 2 in the one health study. For a discussion of the competing theories as to why WTA > WTP readers are referred to Hanemann (1991), Kahneman and Tversky (1979) and Morrison (1998).

Regardless of the reason, the implication is that the threshold dividing the cost-effectiveness plane into cost-effective and not-cost-effective regions is not a straight line through the origin with slope equal to the WTP, as seen in Figure 1.1. The threshold is now made up of two straight lines meeting at the origin. One, for $\Delta_e > 0$, has slope equal to the WTP, and the other, for $\Delta_e < 0$, has slope equal to the WTA. The new 'kinked' threshold is illustrated in Figure 4.9. The darkest shaded region is cost-effective if WTA = WTP, but not if WTA > WTP.

The cost-effectiveness analysis of a clinical trial needs to be adjusted if WTA > WTP. Let WTP $= \lambda$ and let WTA $= \gamma\lambda$, where $\gamma \geq 1$. The parameters Δ_e and Δ_c are estimated in the appropriate way, as discussed in previous chapters and earlier in this chapter, to yield $\hat{\Delta} = (\hat{\Delta}_e, \hat{\Delta}_c)^T$ with estimated covariance matrix $\hat{V}(\hat{\Delta})$. Recalling from Section 4.4

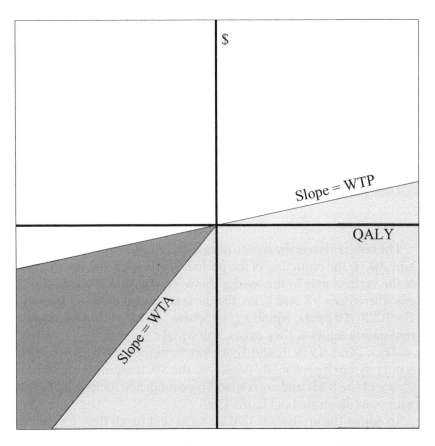

Figure 4.9 threshold for CE plane with WTA > WTP

that $f_{\hat{\Delta}}(\cdot, \cdot)$ is the density function for a bivariate normal with mean $\hat{\Delta}$ and covariance matrix $\hat{V}(\hat{\Delta})$ and that the CEAC for $\gamma = 1$ is given by

$$\mathcal{A}(\lambda) = \int\limits_{-\infty}^{\infty} \int\limits_{-\infty}^{\lambda e} f_{\hat{\Delta}}(e, c) \, dc \, de = \Phi\left(\hat{b}_{\lambda}/\sqrt{v_{\lambda}}\right)$$

where $\hat{b}_{\lambda}$ and v_{λ} are defined as in Section 4.3.

When $\gamma > 1$ the upper limit of the inner integral depends on whether e (i.e. Δ_e) is positive or negative, being λe when $e \geq 0$ and

$\gamma\lambda e$ when $e < 0$. Therefore, to account for $\gamma > 1$ the CEAC becomes

$$\mathcal{A}_\gamma(\lambda) = \int_{-\infty}^{0} \int_{-\infty}^{\gamma\lambda e} f_{\hat{\Delta}}(e, c)\, dc\, de + \int_{0}^{\infty} \int_{-\infty}^{\lambda e} f_{\hat{\Delta}}(e, c)\, dc\, de$$

$$= \mathcal{A}(\lambda) - \int_{-\infty}^{0} \int_{\gamma\lambda e}^{\lambda e} f_{\hat{\Delta}}(e, c)\, dc\, de$$

$$= \Phi\left(\hat{b}_\lambda/\sqrt{v_\lambda}\right) - \int_{-\infty}^{0} \int_{\gamma\lambda e}^{\lambda e} f_{\hat{\Delta}}(e, c)\, dc\, de$$

The last step is seen by recognizing that the difference between $\mathcal{A}_\gamma(\lambda)$ and $\mathcal{A}(\lambda)$ is the reduction of the probability density that lies to the left of the vertical axis in the 'wedge' between the lines through the origin with slopes $\gamma\lambda$ and λ, i.e. the darkest shaded area in Figure 4.9. The ICER, if it exists, equals $r_{\mathcal{A}_\gamma,0.5}$ where $\mathcal{A}_\gamma(r_{\mathcal{A}_\gamma,\beta}) = \beta$ and the corresponding limits, if they exists, will equal $r_{\mathcal{A}_\gamma,\alpha}$ and $r_{\mathcal{A}_\gamma,1-\alpha}$. Clearly $\mathcal{A}_\gamma(\lambda) < \mathcal{A}(\lambda)$, if $\gamma > 1$, and if both functions increase with λ (i.e. there is not too much probability density in the SW quadrant), then the estimate of the ICER and corresponding confidence limits will increase with γ, as illustrated in Figure 4.10.

To adjust the analysis of INB we must first recall that, for the case where WTA = WTP = λ, the estimate of INB and corresponding limits are given, respectively, by $b_{\mathcal{P},0.5}$, $b_{\mathcal{P},\alpha}$ and $b_{\mathcal{P},1-\alpha}$, where $\mathcal{P}(b_{\mathcal{P},\beta}) = \beta$ and

$$\mathcal{P}(b) = \int_{-\infty}^{\infty} \int_{\lambda e-b}^{\infty} f_{\hat{\Delta}}(e, c)\, dc\, de = \Phi\left((b - \hat{b}_\lambda)/\sqrt{v_\lambda}\right) \qquad (4.10)$$

as defined in Equation (4.9). The function $\mathcal{P}(\cdot)$ is being treated as the posterior distribution for INB, i.e. $\mathcal{P}(b) = \Pr(\Delta_e\lambda - \Delta_e \leq b)$, and assuming normality and a flat prior yields Equation (4.10), see Willan, O'Brien and Leyva (2001). When $\gamma > 1$ the lower limit of the inner integral depends on whether e (i.e. Δ_e) is positive or negative, being $\lambda e - b$ when $e \geq 0$ and $\gamma\lambda e - b$ when $e < 0$. Therefore, to account

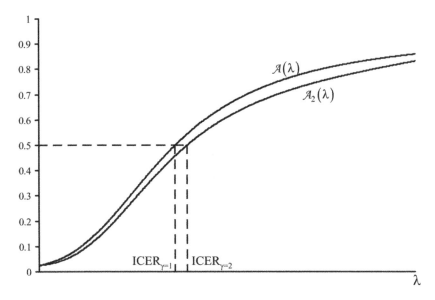

Figure 4.10 CEACs for $\gamma = 1$ and $\gamma = 2$

for $\gamma > 1$ the posterior distribution for INB becomes

$$\mathcal{P}_\gamma(b) = \int\limits_{-\infty}^{0} \int\limits_{\gamma\lambda e-b}^{\infty} f_{\hat{\Delta}}(e,c)\,dc\,de + \int\limits_{0}^{\infty} \int\limits_{\lambda e-b}^{\infty} f_{\hat{\Delta}}(e,c)\,dc\,de$$

$$= \mathcal{P}(b) + \int\limits_{-\infty}^{0} \int\limits_{\gamma\lambda e-b}^{\lambda e-b} f_{\hat{\Delta}}(e,c)\,dc\,de$$

$$= \Phi((b - \hat{b}_\lambda)/\sqrt{v_\lambda}) + \int\limits_{-\infty}^{0} \int\limits_{\gamma\lambda e-b}^{\lambda e-b} f_{\hat{\Delta}}(e,c)\,dc\,de$$

The last step is seen by recognizing that the difference between $\mathcal{P}_\gamma(\lambda)$ and $\mathcal{P}(\lambda)$ is the addition of the probability density that lies to the left of the vertical axis in the 'wedge' between the lines through the point $(0, -b)$ with slopes $\gamma\lambda$ and λ, i.e. the darkest shaded area in

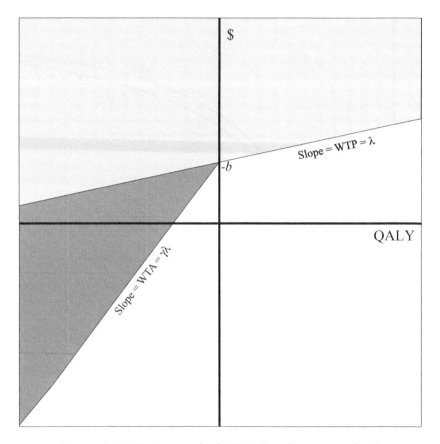

Figure 4.11 Region in which INB is less than or equal to b

Figure 4.11. The estimate of INB and corresponding limits are given by $b_{\mathcal{P}_\gamma, 0.5}$, $b_{\mathcal{P}_\gamma, \alpha}$ and $b_{\mathcal{P}_\gamma, 1-\alpha}$, respectively, where $\mathcal{P}_\gamma(b_{\mathcal{P}_\gamma, \beta}) = \beta$. Clearly $\mathcal{P}_\gamma(\lambda) > \mathcal{P}(\lambda)$, if $\gamma > 1$, and therefore estimate of the INB and corresponding confidence limits will decrease with γ, as illustrated in Figure 4.12.

4.8   SUMMARY

Contained in this chapter are methods of statistical inference for cost-effectiveness trials using estimators of the five key parameters. Note

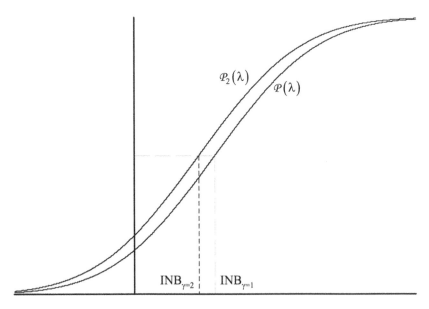

Figure 4.12 Posterior distribution for INB for $\gamma = 1$ and $\gamma = 2$

that the presence of censoring affects only how these five basic parameters are estimated (see Chapter 2 for non-censored data and Chapter 3 for censored data) not how they are subsequently used in analyses. Although the chapter has reviewed in detail both the ICER and INB approach to analyzing cost-effectiveness information and emphasized their formal equivalence, it should be clear that the INB approach offers a number of statistical and practical advantages, avoiding as it does the problems associated with ratio statistics. Nevertheless, in making use of the threshold decision rule, λ, which is generally unknown to the analyst, it is recommended that any analysis be presented as a function of λ. This could be either in terms of the INB and its associated confidence interval directly or as a CEAC which has its most natural interpretation within a Bayesian framework. In the following chapter the methods of estimation and inference reviewed in this chapter are illustrated using several examples.

Cost-effectiveness Analysis: Examples

5.1 INTRODUCTION

Five examples of cost-effectiveness analyses will be given in detail in this chapter, illustrating different aspects of the material covered in preceding chapters. For non-censored data, an example with a binary measure of effectiveness is given in Section 5.2 and an example with quality-adjusted survival time as the measure of effectiveness is given in Section 5.3. For censored data, an example with survival time as the measure of effectiveness is given in Section 5.4 and an example with quality-adjusted survival time as the measure of effectiveness is given in Section 5.6. A final example presents a Bayesian analysis contrasting the use of prior information and uninformative priors. Emphasis will be placed on using the appropriate methodology for estimating the five parameters of interest, and on using the plot of the incremental net benefit by willingness-to-pay and the cost-effectiveness acceptability curve for inference. The use of different decision thresholds in the SW and NE quadrants of the cost-effectiveness plane is also demonstrated.

5.2 THE CADET-Hp TRIAL

The CADET-Hp Trial is a double-blind, placebo-controlled, parallel-group, multicenter, randomized controlled trial performed in 36

Statistical Analysis of Cost-effectiveness Data. A. Willan and A. Briggs
© 2006 John Wiley & Sons, Ltd.

family practitioner centres across Canada. The results are published in Chiba *et al.* (2002) and Willan (2004). Patients 18 years and over with uninvestigated dyspepsia of at least moderate severity presenting to their family physicians were eligible for randomization, provided they did not have any alarm symptoms and were eligible for empiric drug therapy. Patients were randomized between

> *T*: Omeprazole 20 mg, metronidazole 500 mg and clarithromycin 250 mg, and
> *S*: Omeprazole 20 mg, placebo metronidazole and placebo clarithromycin.

Both regimens were given twice daily for seven days. Treatment success was defined as the presence of no or minimal dyspepsia symptoms at one year. Total costs were determined from the societal perspective and are given in Canadian dollars.

A summary of the parameter estimates is given in Table 5.1. The elements in Table 5.1 are calculated by applying Equations (2.1–2.5). Costs are given in Canadian dollars (CAD\$). *T* is observed to increase success and decrease costs and, consequently, the point $(\hat{\Delta}_e, \hat{\Delta}_c)$ lies in the SE (win–win) quadrant and $\hat{b}_\lambda$ is positive for all positive values of λ. For sake of argument suppose the WTP for a success is CAD\$ 1000. That is, if the true value of INB evaluated at 1000 is less than or equal to zero, *T* should not be adopted. This, then, becomes the null hypothesis, expressed below as H_0. On the other hand, if INB evaluated at 1000 is greater than zero, *T* should be adopted. This becomes the

Table 5.1 Parameter estimates for the CADET-Hp trial

	$T(n_T = 142)$	$S(n_S = 146)$		Equation
$\hat{\pi}_j$	0.5070	0.3699	difference $= \hat{\Delta}_e = 0.1371$	(2.3)
$\hat{v}_j$	476.97	529.98	difference $= \hat{\Delta}_c = -53.01$	(2.1)
$\hat{V}(\hat{\pi}_j)$	0.00176	0.001596	sum $= \hat{V}(\hat{\Delta}_e)$ $= 0.003356$	(2.4)
$\hat{V}(\hat{v}_j)$	2167	2625	sum $= \hat{V}(\hat{\Delta}_c) = 4792$	(2.2)
$\hat{C}(\hat{\pi}_j, \hat{v}_j)$	−0.2963	−0.4166	sum $= \hat{C}(\hat{\Delta}_e, \hat{\Delta}_c)$ $= -0.7129$	(2.5)

alternative hypothesis, expressed below as H_1. Thus the investigators wish to test the hypothesis

$$H_0: b_{1000} \leq 0 \quad \text{versus} \quad H_1: b_{1000} > 0$$

at the 5% level. Rejection of H_0 in favor of H_1 provides evidence that T is cost-effective compared with S. First, the investigators must determine:

$$\hat{b}_{1000} = 1000\hat{\Delta}_e - \hat{\Delta}_c = 1000 \times 0.1371 - (-53.01) = 190.1$$

and

$$\begin{aligned} v_{1000} &= \hat{V}\left(\hat{b}_{1000}\right) = 1000^2 \times \hat{V}\left(\hat{\Delta}_e\right) + \hat{V}\left(\hat{\Delta}_c\right) - 2 \times 1000 \\ &\quad \times C\left(\hat{\Delta}_e, \hat{\Delta}_c\right) = 1000^2 \times 0.003356 + 4792 - 2 \times 1000 \\ &\quad \times (-0.7129) = 9574 \end{aligned}$$

The z-statistic for testing H_0 is $\hat{b}_{1000}/\sqrt{v_{1000}} = 190.1/\sqrt{9574} = 1.943$. Since $1.943 > z_{0.95} (= 1.645)$, H_0 can be rejected in favour of H_1, and the investigators can conclude that there is evidence that T is cost-effective compared with S if the WTP is at least CAD\$ 1000. The 90% confidence limits for b_{1000} are given by:

$$\hat{b}_{1000} \pm z_{0.95}\sqrt{v_{1000}} = 190.1 \pm 1.645 \times 97.85 = 29.15, 351.1.$$

Since the lower limit is greater than 0, we can, as above, reject H_0 in favour of H_1, at the 5% level. A two-sided 90% confidence interval is constructed because we wish to test a one-sided null hypothesis at the 5% level.

For any value of λ, the quantities $\hat{b}_\lambda$ and v_λ are given by $0.1371\lambda + 53.01$ and $0.003356\lambda^2 + 4792 + 2 \times 0.7129\lambda$, respectively, and the 90% confidence limits are given by $0.1371\lambda + 53.01 \pm 1.645\sqrt{0.003356\lambda^2 + 4792 + 2 \times 0.7129\lambda}$. The quantity $\hat{b}_\lambda$ and corresponding 90% confidence limits can be calculated for a large range of λ and plotted as shown in Figure 5.1. The plot of $\hat{b}_\lambda$ has slope $0.1371 (= \Delta_e)$, crosses the vertical axis at $53.01 (= -\hat{\Delta}_c)$ and the horizontal axis at $-53.01/0.1371 = -386.7 (= \text{ICER})$. By observing where the confidence limits cross the vertical axis, one can make inference regarding the difference between arms with respect to cost, in what is essentially a cost-minimization analysis. Since the

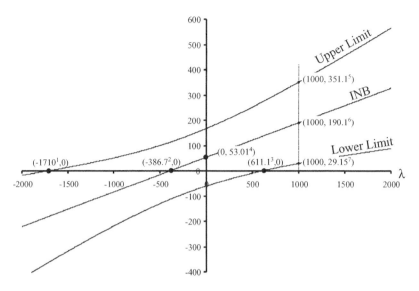

Figure 5.1 INB versus λ for the CADET-Hp trial. [1]Lower confidence limit for ICER; [2]ICER; [3]upper confidence limit for ICER; [4]estimate of $-\Delta_c$; [5]upper confidence limit for INB(1000); [6]estimate of INB(1000); [7]lower confidence limit for INB(1000)

confidence interval includes 0, the null hypothesis of no difference in mean cost cannot be rejected at the 10% level (note that this is a two-sided test). The horizontal intercepts of the confidence limits for b_λ define the Fieller confidence interval for the ICER. Therefore, by focusing attention on the horizontal axis one can perform a cost-effectiveness analysis using an ICER approach. The 90% confidence limits for the ICER are -1710 and 611.1. The hypothesis that the ICER is greater than any value above 611.1 can be rejected at the 5% level since 611.1 is the upper limit of the 90% confidence interval. The Fieller 90% confidence limits for the ICER can be calculated directly from applying Equation (4.4). That is, the 90% confidence limits are given by:

$$\hat{R}\left\{\left(\left(1 - z_{1-0.05}^2 c \pm z_{1-0.05}\sqrt{a + b - 2c - z_{1-0.05}^2(ab - c^2)}\right)\right) \middle/ \left(1 - z_{1-0.05}^2 a\right)\right\}$$

Table 5.2   Elements for calculation confidence limits for ICER for the CADET-Hp trial

$\hat{R} = \hat{\Delta}_c/\hat{\Delta}_e = -53.01/0.1371 = -386.7$

$a = \hat{V}(\hat{\Delta}_e)/\hat{\Delta}_e^2 = 0.003356/0.1371^2 = 0.1785$

$b = \hat{V}(\hat{\Delta}_c)/\hat{\Delta}_c^2 = 4792/(-53.01)^2 = 1.705$

$c = \hat{C}(\hat{\Delta}_e, \hat{\Delta}_c)/(\hat{\Delta}_e\hat{\Delta}_c) = -0.7129/(0.1371 \times (-53.01)) = 0.09809$

$z_{1-0.05} = 1.645$

The associated elements for Equation (4.4) are given in Table 5.2. Substituting them into the above equation yields the limits -1710 and 611.1.

Using Figure 5.1 one can perform a cost-effectiveness analysis using an INB approach for any value of λ. In particular, for $\lambda = 1000$, the confidence interval contains only positive values, which means, as above, that the null hypothesis of $b_{1000} \leq 0$ can be rejected at the 5% level, and that there is evidence to support the adoption of T. This is true for any value of λ that is greater than 611.1 (the upper limit of the ICER). For values of λ less than 611.1 the confidence interval includes negative values and the null hypothesis that $b_\lambda \leq 0$ cannot be rejected at the 5% level.

The cost-effectiveness acceptability curve (CEAC) is given by

$$A(\lambda) = \Phi\left(b(\lambda)\bigg/\sqrt{\hat{V}(b(\lambda))}\right)$$

$$= \Phi\left((0.1371\lambda + 53.01)\bigg/\sqrt{0.003356\lambda^2 + 4792 + 2 \times 0.7129\lambda}\right)$$

where $\Phi(\cdot)$ is the cumulative distribution function for a standard normal random variable, and can be calculated using the following simple SAS^{TM} code:

```
do lambda = 0 to 1000;
      a = probnorm((0.1372*lambda + 53.01)
              /sqrt(0.003356*lambda**2 + 4792
                  + 2*0.7129*lambda));
      output;
end;
```

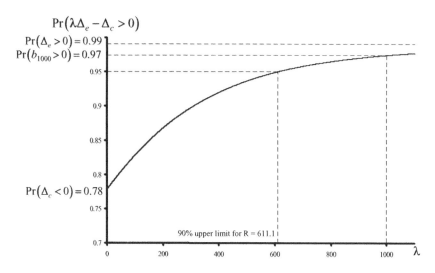

Figure 5.2 Cost-effectiveness acceptability curve for CADET-Hp trial

The CEAC is shown in Figure 5.2. Since the ICER and its lower 90% confidence limit are negative, they are not shown in the figure. The curve meets the vertical axis ($\lambda = 0$) at $\mathcal{A}(0) = \Phi(\hat{b}_0/\sqrt{v_0}) = \Phi(-\hat{\Delta}_c/\sqrt{\hat{V}(\hat{\Delta}_c)}) = \Pr(\Delta_c < 0) = \Phi(53.01/\sqrt{4792}) = 0.7781$ and asymptotically approaches $\Pr(\Delta_e > 0) = \Phi(\hat{\Delta}_e/\sqrt{\hat{V}(\hat{\Delta}_e)}) = \Phi(0.1371/\sqrt{0.003356}) = 0.9910$. Since $\mathcal{A}(1000) = 0.9741$, if the WTP for a success is CAD\$ 1000, we could say that the probability that T is cost-effective is 97.41%.

5.3 SYMPTOMATIC HORMONE-RESISTANT PROSTATE CANCER

In a trial of symptomatic, hormone-resistant prostate cancer, 161 patients were randomized between prednisone alone (S) and prednisone plus mitoxantrone (T). The clinical results are published in Tannock *et al.* (1996) and the economics analysis in Bloomfield *et al.* (1998). Although there was no statistically significant difference in survival, patients experienced a better quality of life with T. Cost data,

Table 5.3 Parameter estimates for the prostate trial

	$T(n_T = 61)$	$S(n_S = 53)$		Equation
$\hat{\varphi}_j$	40.89	28.11	difference $= \hat{\Delta}_e = 12.78$	(2.10)
$\hat{v}_j$	27\,322	29\,039	difference $= \hat{\Delta}_c = -1717$	(2.1)
$\hat{V}(\hat{\varphi}_j)$	24.10	16.42	sum $= \hat{V}(\hat{\Delta}_e) = 40.52$	(2.11)
$\hat{V}(\hat{v}_j)$	6466\,351	7872\,681	sum $= \hat{V}(\hat{\Delta}_c) = 14\,339\,032$	(2.2)
$\hat{C}(\hat{\varphi}_j, \hat{v}_j)$	2771	2876	sum $= \hat{C}(\hat{\Delta}_e, \hat{\Delta}_c) = 5647$	(2.12)

including hospital admissions, outpatient visits, investigations, therapies and palliative care, were collected retrospectively on the 114 patients from the three largest centres. Survival was quality-adjusted using the EORTC quality of life questionnaire QLQ-C30. All patients were followed until death.

A summary of the parameter estimates is given in Table 5.3. Costs are given in Canadian dollars and effectiveness in quality-adjusted life-weeks (QALW). T is observed to increase quality-adjusted survival and decrease costs, and consequently, the point $(\hat{\Delta}_e, \hat{\Delta}_c)$ lies in the SE (win–win) quadrant and $\hat{b}_\lambda$ is positive for all positive values of λ. If the WTP for a QALW is CAD\$ 500, the investigators wish to test the hypothesis

$$H_0 : b_{500} \leq 0 \quad \text{versus} \quad H_1 : b_{500} > 0$$

at the 5% level, and conclude that there is evidence to adopt T if H_0 is rejected. First, the investigators must determine:

$$\hat{b}_{500} = 500 \times \hat{\Delta}_e - \hat{\Delta}_c = 500 \times 12.78 - (-1717) = 8107$$

and

$$
\begin{aligned}
v_{500} &= 500^2 \times \hat{V}\left(\hat{\Delta}_e\right) + \hat{V}\left(\hat{\Delta}_c\right) - 2 \times 500 \times C\left(\hat{\Delta}_e, \hat{\Delta}_c\right) \\
&= 500^2 \times 40.52 + 14\,339\,032 - 2 \times 500 \times 5647 \\
&= 18\,822\,032
\end{aligned}
$$

The z-statistic for testing H_0 is $\hat{b}_{500}/\sqrt{v_{500}} = 8107/\sqrt{18\,822\,032} = 1.869$. Since $1.869 > z_{0.95}(= 1.645)$, H_0 can be rejected in favour of

H_1, and the investigators can conclude that there is evidence that T is cost-effective compared with S if the WTP is at least CAD\$ 500. The 90% confidence limits for b_{500} are given by:

$$\hat{b}_{500} \pm z_{0.95}\sqrt{v_{500}} = 8107 \pm 1.645 \times 4338 = 971.0, 15\,243$$

Since the lower limit is greater than 0, we can, as above, reject H_0 in favour of H_1, at the 5% level. A two-sided 90% confidence interval is constructed because we wish to test a one-sided null hypothesis at the 5% level.

For any value of λ, the quantities $\hat{b}_\lambda$ and v_λ are given by $12.78\lambda + 1717$ and $40.52\lambda^2 + 14\,339\,032 - 2 \times 5647\lambda$, respectively, and the 90% confidence limits are given by $12.78\lambda + 1717 \pm 1.645 \times \sqrt{40.52\lambda^2 + 14\,339\,032 - 2 \times 5647\lambda}$. The quantity $\hat{b}_\lambda$ and corresponding 90% confidence limits can be calculated for a large range of λ and plotted as shown in Figure 5.3. The plot of $\hat{b}_\lambda$ has slope 12.78

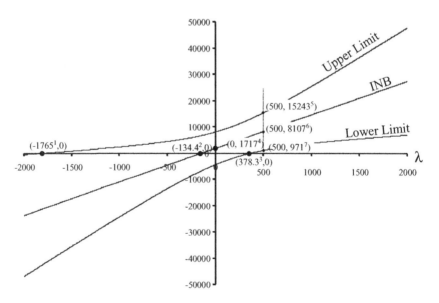

Figure 5.3 INB versus λ for the prostate trial. [1]Lower confidence limit for ICER; [2]ICER; [3]upper confidence limit for ICER; [4]estimate of $-\Delta_c$; [5]upper confidence limit for INB(500); [6]estimate of INB(500); [7]lower confidence limit for INB(500)

$(= \Delta_e)$, crosses the vertical axis at $1717 (= -\hat{\Delta}_c)$ and the horizontal axis at $-1717/12.78 = -134.4 (= \text{ICER})$. By observing where the confidence limits cross the vertical axis, one can make inference regarding the difference between arms with respect to cost, in what is essentially a cost-minimization analysis. Since the confidence interval includes 0, the null hypothesis of no difference in mean cost cannot be rejected at the 10% level (note that this is a two-sided test). The horizontal intercepts of the confidence limits define the Fieller confidence interval for the ICER. Therefore, by focusing attention on the horizontal axis one can perform a cost-effectiveness analysis using an ICER approach. The 90% confidence limits for the ICER are -1765 and 378.3. The hypothesis that the ICER is greater than any value above 378.3 can be rejected at the 5% level since 378.3 is the upper limit of the 90% confidence interval. The 90% confidence limits for the ICER can be calculated directly from applying Equation (4.4). That is, 90% confidence limits are given by:

$$\hat{R}\left\{\left(1 - z_{1-0.05}^2 c \pm z_{1-0.05}\sqrt{a + b - 2c - z_{1-0.05}^2(ab - c^2)}\right) \middle/ \left(1 - z_{1-0.05}^2 a\right)\right\}$$

The associated elements for Equation (4.4) are given in Table 5.4.

Using Figure 5.3 one can perform a cost-effectiveness analysis using an INB approach for any value of λ. In particular, for $\lambda = 500$, the confidence interval contains only positive values, which means, as above, the null hypothesis that $b_{500} \leq 0$ can be rejected and that there is evidence to support the adoption of T. This is true for any value of

Table 5.4 Elements for calculation confidence limits for ICER for the prostate trial

---

$\hat{R} = \hat{\Delta}_c/\hat{\Delta}_e = -1717/12.78 = -134.4$

$a = \hat{V}(\hat{\Delta}_e)/\hat{\Delta}_e^2 = 40.52/12.78^2 = 0.2481$

$b = \hat{V}(\hat{\Delta}_c)/\hat{\Delta}_c^2 = 14\,339\,032/(-1717)^2 = 4.864$

$c = \hat{C}(\hat{\Delta}_e, \hat{\Delta}_c)/(\hat{\Delta}_e\hat{\Delta}_c) = 5647/(12.78 \times (-1717)) = -0.2573$

$z_{1-0.05} = 1.645$

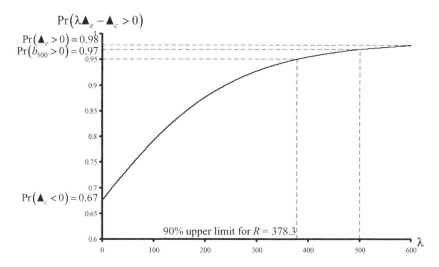

Figure 5.4 Cost-effectiveness acceptability curve for prostate trial

λ that is greater than 378.3 (upper limit of the ICER). For values of λ less than 378.3 the confidence interval includes negative values and the null hypothesis that $b_\lambda \leq 0$ cannot be rejected.

The cost-effectiveness acceptability curve can be calculated by:

$$\mathcal{A}(\lambda) = \Phi\left(\hat{b}_\lambda \sqrt{v_\lambda}\right) = \Phi\left((12.78\lambda + 1717)/\right.$$
$$\left. \sqrt{40.52\lambda^2 + 14\,339\,032 - 2 \times 5647\lambda}\right)$$

The curve is illustrated in Figure 5.4 and meets the vertical axis at $\Pr(\Delta_c < 0) = \Phi(1717/\sqrt{14\,339\,032}) = 0.6749$ and asymptotically approaches $\Pr(\Delta_e > 0) = \Phi(12.78/\sqrt{40.52}) = 0.9777$. Since $\mathcal{A}(500) = 0.9692$, if the WTP for a QALW is CAD\$500, we could say that the probability that T is cost-effective is 96.92%.

The potential use of different threshold decision rules in the SW and NE quadrants of the cost-effectiveness plane is demonstrated for the prostate example in Table 5.5. In the upper panel of the table, the probability that Treatment is cost-effect and the estimate of INB, with 90% Bayesian credible intervals are presented under the assumption of a WTP of CAD\$400 per quality-adjusted life-week, for various values of γ, the ratio of WTP to WTA in the two quadrants (see Section 4.6).

Table 5.5   Estimate of INB and Bayesian interval for $= 400/\text{QALW}$

	γ	$\mathcal{A}_\gamma(400)$	INB	90% credible interval
$\hat{\Delta}_e = 12.8$	1	0.9546	6836	195–13 477
	2	0.9526	6828	116–13 477
	5	0.9475	6817	−92–13 476
	10	0.9434	6811	−281–13 476
$\hat{\Delta}_e = 0$	1	0.6647	1715	−4925–691
	2	0.5759	917	−7215–6629
	5	0.4727	−448	−18 679–6409
	10	0.4324	−1515	−39 146–6328

The first row with $\gamma = 1$ corresponds to the conventional approach of using the same value of lambda throughout. Since the estimate $\hat{\Delta}_e$ is statistically significant, and little of the probability density of $(\hat{\Delta}_e, \hat{\Delta}_c)^T$ lies to the left of the vertical axis, as γ increases, only small changes are seen. However, the hypothesis $b_{400} \leq 0$ (versus $b_{400} > 0$) can be rejected for $\gamma = 1$ or 2, but not for $\gamma = 5$ or 10. For illustration, the analysis was rerun setting $\hat{\Delta}_e = 0$, so that 50% of the probability lies to the left of the vertical axis. The results, shown in the bottom panel in Table 5.5, illustrate that in this circumstance the results are quite sensitive to γ.

5.4   THE CANADIAN IMPLANTABLE DEFIBRILLATOR STUDY (CIDS)

In a trial of patients at risk of cardiac arrest, a total of 659 patients with resuscitated ventricular fibrillation or sustained ventricular tachycardia or with unmonitored syncope were randomized between amiodarone and implantable cardioverter defibrillator. Due to budgetary constraints, the costs were collected on the first 430 patients only. The clinical results are reported in Connolly *et al.* (2000), and the economic evaluation in O'Brien *et al.* (2000). For this example the measure of effectiveness is survival time, with the duration of interest set to 6.5 years, i.e. $\tau = 6.5$. Administrative censoring occurred and,

Table 5.6 Parameter estimates for the CIDS trial using inverse probability weighting

	$T(n_T = 212)$	$S(n_S = 218)$		Equation
$\hat{\mu}_j$	4.832	4.682	difference $= \hat{\Delta}_e = 0.1500$	(3.27)
$\hat{v}_j$	87 044	38 819	difference $= \hat{\Delta}_c = 48\,225$	(3.12, 3.13)
$\hat{V}(\hat{\mu}_j)$	0.02418	0.02440	sum $= \hat{V}(\hat{\Delta}_e) = 0.04858$	(3.28)
$\hat{V}(\hat{v}_j)$	8462 152	6497 962	sum $= \hat{V}(\hat{\Delta}_c)$ $= 14\,960\,114$	(3.14)
$\hat{C}(\hat{\mu}_j, \hat{v}_j)$	125.8	20.42	sum $= \hat{C}(\hat{\Delta}_e, \hat{\Delta}_c)$ $= 146.2$	(3.30)

consequently, not all patients were followed for the entire duration of interest. Total costs are given in Canadian dollars, and were collected every three months for a total of 26 three-monthly intervals.

The estimators of the parameters of interest, using inverse probability weighting (IPW), are given in Table 5.6. By contrast the parameters of interest using life-table methods for effectiveness and the direct method for cost are given in Table 5.7. The estimates are almost identical and the following analyses are performed using the IPW estimates. T is observed to increase mean survival and increase costs, and consequently, the point $(\hat{\Delta}_e, \hat{\Delta}_c)$ lies in the NE (trade-off) quadrant. For any value of λ, the quantities $\hat{b}_\lambda$ and v_λ are given

Table 5.7 Parameter estimates for the CIDS trial using life-table methods for effectiveness and the direct method for cost

	$T(n_T = 212)$	$S(n_S = 218)$		Equation
$\hat{\mu}_j$	4.832	4.682	difference $= \hat{\Delta}_e = 0.1500$	(3.21)
$\hat{v}_j$	87 103	38 864	difference $= \hat{\Delta}_c = 48\,239$	(3.2)
$\hat{V}(\hat{\mu}_j)$	0.02443	0.02437	sum $= \hat{V}(\hat{\Delta}_e) = 0.04858$	(3.22)
$\hat{V}(\hat{v}_j)$	8461 538	6519 142	sum $= \hat{V}(\hat{\Delta}_c)$ $= 14\,980\,680$	(3.5)
$\hat{C}(\hat{\mu}_j, \hat{v}_j)$	124.9	14.20	sum $= \hat{C}(\hat{\Delta}_e, \hat{\Delta}_c)$ $= 139.1$	(3.26)

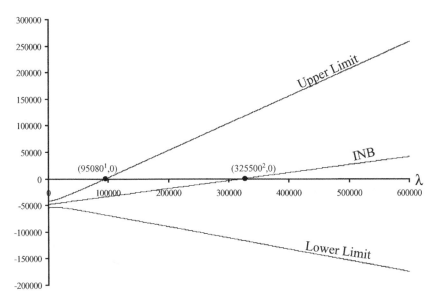

Figure 5.5 INB versus λ for the CIDS trial. [1]Lower confidence limit for ICER; [2]ICER

by $0.15\lambda - 48225$ and $0.04858\lambda^2 + 14\,960\,114 - 2 \times 146.2\lambda$, respectively, and the 90% confidence limits are given by $0.15\lambda - 48\,225 \pm 1.645 \times \sqrt{0.04858\lambda^2 + 14\,960\,114 - 2 \times 146.2\lambda}$. The quantity $\hat{b}_\lambda$ and corresponding 90% confidence limits are plotted in Figure 5.5. The plot of $\hat{b}_\lambda$ has slope $0.15 \ (= \Delta_e)$, crosses the vertical axis at $-48225 \ (= -\hat{\Delta}_c)$ and the horizontal axis at $48\,225/0.15 = 321\,500 \ (= \text{ICER})$. By observing where the confidence limits cross the vertical axis, one can make inference regarding the difference between arms with respect to cost, in what is essentially a cost-minimization analysis. Since the confidence interval includes only negative values, the null hypothesis of no difference in mean cost can be rejected at the 10% level, and the investigators can conclude that there is evidence that T increases costs. (Recall that the confidence interval defined on the vertical axis is for $-\Delta_c$, and since the confidence interval for $-\Delta_c$ includes only negative values, the confidence interval for Δ_c includes only positive values.)

Table 5.8   Elements for calculation confidence limits for ICER for the CIDS trial

---

$\hat{R} = \hat{\Delta}_c/\hat{\Delta}_e = 48\,225/0.1500 = 321\,500$

$a = \hat{V}(\hat{\Delta}_e)/\hat{\Delta}_e^2 = 0.04858/0.1500^2 = 2.159$

$b = \hat{V}(\hat{\Delta}_c)/\hat{\Delta}_c^2 = 14\,960\,114/(48\,225)^2 = 0.006433$

$c = \hat{C}(\hat{\Delta}_e, \hat{\Delta}_c)/(\hat{\Delta}_e\hat{\Delta}_c) = 146.2/(0.1500 \times 48\,225) = 0.02021$

$z_{1-0.05} = 1.645$

The horizontal intercepts of the confidence limits define the Fieller confidence interval for the ICER and, by focusing attention on the horizontal axis, one can perform a cost-effectiveness analysis using an ICER approach. The 90% lower limit for the ICER is 95 080. Since the lower limit of the INB fails to cross the horizontal axis, the upper limit of the ICER approaches $+\infty$, i.e. the confidence interval for the ICER includes the positive vertical axis of the cost-effectiveness plane. This is interpreted to mean that there is no value for the WTP that would make T cost-effective with 95% confidence. Likewise, since the lower limit of the INB fails to cross the horizontal axis, the confidence interval for INB always includes negative values, which leads to the same conclusion. The 90% confidence limits for the ICER can be calculated directly from applying Equation (4.4), whose associated elements are given in Table 5.8. The solution to Equation (4.4) yields the interval [95 080, −220 626]. Since the lower limit is positive and the upper limit is negative, the confidence region on the CE plane includes part of the NW (lose–lose) quadrant, see Figure 5.6. Consequently, a proper representation of the interval is given by [95 080, $+\infty$), since the largest value in the interval approaches $+\infty$.

The cost-effectiveness acceptability curve can be calculated by:

$$A(\lambda) = \Phi\left(\hat{b}_\lambda/\sqrt{v_\lambda}\right)$$
$$= \Phi\left((0.15\lambda - 48225)\big/\sqrt{0.04858\lambda^2 + 14960114 - 2 \times 146.2\lambda}\right)$$

The curve, see Figure 5.7, meets the vertical axis ($\lambda = 0$) at $\Pr(\Delta_c < 0) = \Phi(-48\,225/\sqrt{14\,960\,114}) = 5.564 \times 10^{-36}$ and asymptotically approaches $\Pr(\Delta_e > 0) = \Phi(0.15/\sqrt{0.04858}) = 0.7519$.

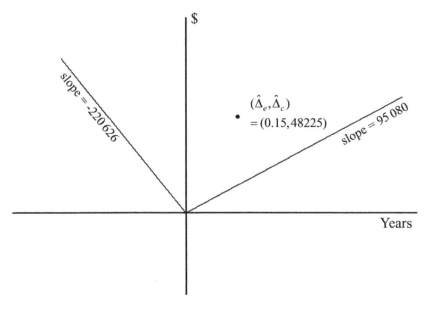

Figure 5.6 ICER confidence region on the CE plane for the CIDS trial

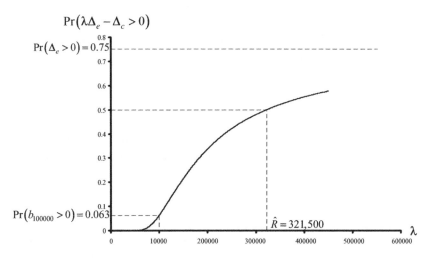

Figure 5.7 Cost-effectiveness acceptability curve for CIDS trial

Since $\mathcal{A}(100\,000) = 0.06314$, if the WTP for a year of life is CAD\$100,000, we could say that the probability that T is cost-effective is 6.314%.

5.5   THE EVALUATE TRIAL

The EVALUATE trial is a randomized comparison of laparoscopic-assisted hysterectomy (T) versus standard hysterectomy (S). Detailed reports of the trial can be found elsewhere, see Garry *et al.* (2004) and Sculpher *et al.* (2004). Patients were stratified into two groups, those undergoing vaginal hysterectomy and those undergoing abdominal hysterectomy, and the strata were analyzed and reported separately. Randomization was 2-to-1 in favor of laparoscopy. Allocation by treatment group and type of hysterectomy is given in Table 5.9. Effectiveness was expressed in quality-adjusted life-years, and measured using the EQ-5D questionnaire, see Kind (1996). EQ-5D measurements were taken at randomization, and at three follow-up visits: six weeks, four months and one year. These follow-up visits defined the three time intervals (i.e. $K = 3$) with $a_1 = 0, a_2 = 6, a_3 = 17$ and $a_4 = 52$, expressed in weeks. Since the boundaries of the time interval coincided with the EQ-5D measurements (as they often would), q_k, the quality-adjusted life-year score for interval k, equals

$$\frac{(Q_{k+1} + Q_k)}{2} \times \frac{(a_{k+1} - a_k)}{52}$$

where Q_k is the EQ-5D measurement at a_k. Total costs, expressed in UK pounds, were collected for these time intervals also. Although there were no deaths, several patients failed to compete all follow-up visits and were considered censored.

Table 5.9   Treatment by type of hysterectomy, EVALUATE example

	Laparoscopic	Standard
Vaginal ($N = 487$)	324	163
Abdominal ($N = 859$)	573	286

Table 5.10 Parameter estimates for the abdominal patients in the EVALUATE trial using inverse probability weighting

	$T(n_T = 573)$	$S(n_S = 286)$		Equation
$\hat{\varphi}_j$	0.8703	0.8617	difference $= \hat{\Delta}_e$ $= 0.009148$	(3.35)
$\hat{v}_j$	1705.4	1519.6	difference $= \hat{\Delta}_c = 185.8$	(3.12, 3.13)
$\hat{V}(\hat{\varphi}_j)$	0.00003242	0.00007121	sum $= \hat{V}(\hat{\Delta}_e)$ $= 0.0001036$	(3.37)
$\hat{V}(\hat{v}_j)$	3245	7099	sum $= \hat{V}(\hat{\Delta}_c) = 10\,344$	(3.14)
$\hat{C}(\hat{\varphi}_j, \hat{v}_j)$	−0.05912	−0.1748	sum $= \hat{C}(\hat{\Delta}_e, \hat{\Delta}_c)$ $= -0.2339$	(3.39)

The estimators of the parameters of interest for the abdominal patients, using inverse probability weighting, are given in Table 5.10. T is observed to increase mean quality-adjusted survival and increase costs and, consequently, the point $(\hat{\Delta}_e, \hat{\Delta}_c)$ lies in the NE (trade-off) quadrant. For any value of λ, the quantities $\hat{b}_\lambda$ and v_λ are given by $0.009148\lambda - 185.8$ and $0.0001036\lambda^2 + 10\,344 - 2 \times (-0.2339)\lambda$, respectively, and the 90% confidence limits are given by

$$0.009148\lambda - 185.8 \pm 1.645\sqrt{0.0001036\lambda^2 + 10\,344 + 2 \times 0.2339\lambda}$$

The quantity $\hat{b}_\lambda$ and corresponding 90% confidence limits are plotted in Figure 5.8. The plot of $\hat{b}_\lambda$ has slope 0.009148 $(= \Delta_e)$, crosses the vertical axis at -185.8 $(= -\hat{\Delta}_c)$ and the horizontal axis at $185.8/0.009148 = 20\,310$ $(= \text{ICER})$. By observing where the confidence limits cross the vertical axis, one can make inference regarding the difference between arms with respect to cost, in what is essentially a cost-minimization analysis. Since the confidence interval includes only negative values, the null hypothesis of no difference in mean cost can be rejected at the 10% level, and the investigators can conclude that there is evidence that T increases costs. (Recall that the confidence interval defined on the vertical axis is for $-\hat{\Delta}_c$, and since the

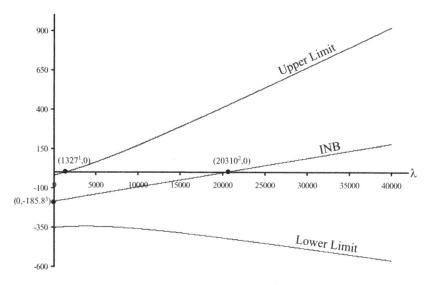

Figure 5.8 INB versus λ for the EVALUATE trial. [1] Lower confidence limit for ICER; [2] ICER; [3] $-\hat{\Delta}_c$

confidence interval for $-\hat{\Delta}_c$ includes only negative values, the confidence interval for $\hat{\Delta}_c$ includes only positive values.)

The horizontal intercepts of the confidence limits define the Fieller confidence interval for the ICER and, by focusing attention on the horizontal axis, one can perform a cost-effectiveness analysis using an ICER approach. The 90% lower limit for the ICER is 1327. Since the lower limit of the INB fails to cross the horizontal axis, the upper limit of the ICER approaches $+\infty$, i.e. the confidence interval for the ICER includes the positive vertical axis of the cost-effectiveness plane. This is interpreted to mean that there is no value for the WTP that would make T cost-effective with 95% confidence. Similarly, since the lower limit of the INB fails to cross the horizontal axis, the confidence interval for INB always includes negative values, which leads to the same conclusion. The 90% confidence limits for the ICER can be calculated directly from applying Equation (4.4), whose associated elements are given in Table 5.11. The solution to Equation (4.4) yields the interval [1327, $-25\,054$]. Since the lower limit is positive and the upper limit is negative, the confidence region on the CE plane includes

Table 5.11   Element for calculation confidence limits for ICER for the abdominal patients in the EVALUATE trial

---

$\hat{R} = \hat{\Delta}_c/\hat{\Delta}_e = 185.8/0.009\,148 = 20\,310$

$a = \hat{V}(\hat{\Delta}_e)/\hat{\Delta}_e^2 = 0.0001036/0.009\,148^2 = 1.238$

$b = \hat{V}(\hat{\Delta}_c)/\hat{\Delta}_c^2 = 10\,344/(185.8)^2 = 0.2996$

$c = \hat{C}(\hat{\Delta}_e, \hat{\Delta}_c)/(\hat{\Delta}_e\hat{\Delta}_c) = -0.2329/(0.009\,148 \times 185.8) = -0.1370$

$z_{1-0.05} = 1.645$

---

part of the NW (lose–lose) quadrant, see Figure 5.9. Consequently, a proper representation of the interval is given by $[1327, +\infty)$, since the largest value in the interval approaches $+\infty$.

The cost-effectiveness acceptability curve can be calculated by

$$A(\lambda) = \Phi\left(\hat{b}_\lambda/\sqrt{v_\lambda}\right)$$

$$= \Phi\left((0.009148\lambda - 185.8)\Big/\sqrt{0.0001036\lambda^2 + 10\,344 + 2 \times 0.2339\lambda}\right)$$

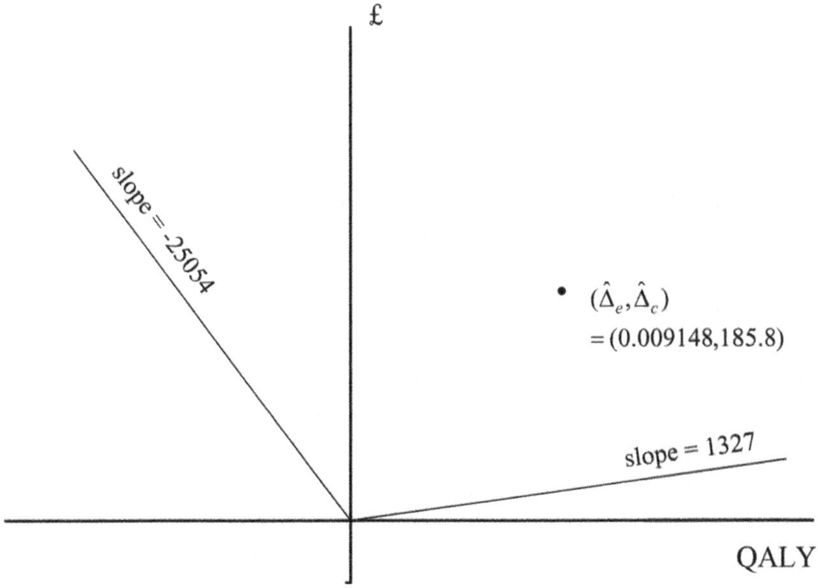

Figure 5.9 ICER confidence region on the CE plane for the EVALUATE trial

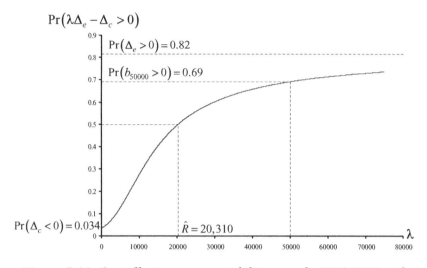

Figure 5.10 Cost-effectiveness acceptability curve for EVALUATE trial

The curve, see Figure 5.10, meets the vertical axis at $\Pr(\Delta_c < 0)$ $= \Phi(-185.8/\sqrt{103\,444}) = 0.03386$ and asymptotically approaches $\Pr(\Delta_e > 0) = \Phi(0.009148/\sqrt{0.0001036}) = 0.8156$. Since $\mathcal{A}(50\,000) = 0.6922$, if the WTP for a year of life is CAD\$ 50 000, we could say that the probability that T is cost-effective is 69.22%.

5.6 BAYESIAN APPROACH APPLIED TO THE UK PDS STUDY

To illustrate the Bayesian approach to cost-effectiveness analysis, data from the UK prospective diabetes study (PDS) are employed. These data relate to a cost-effectiveness analysis designed to assess the efficiency of tight blood pressure control compared with less tight control in hypertensive patients with type 2 diabetes (Raikou *et al.*, 1998). Hypertension in subjects with type II diabetes is a risk factor for macrovascular complications. Although improved blood pressure control has been shown to reduce myocardial infarction and stroke in a diabetic sub-group of elderly patients with type II diabetes (SHEP Cooperative Research Group 1991) no information on younger patients or the effect on complications of diabetes was available. The authors of the original cost-effectiveness analysis noted that, although cost-effectiveness of antihypertensive programs based on education and drugs have been reported for a number of populations: '. . . these analyses have mainly been based on models and lack information on effectiveness and use of resources from long term trials, and none has considered hypertensive patients with type II diabetes.'

The authors then go on to report the results of their study in the form of the incremental costs, incremental effects and employing the cost-effectiveness acceptability curve approach. These results, assuming the then standard discount rates of 6% for both costs and life-years, are reproduced in Table 5.12. Also included in Table 5.12 is an assessment of the net benefit of tight blood pressure control, assuming a value of λ of £20 000 per year of life gained (LYG). Although tight blood pressure control looks both more effective (in terms of LYG) and more costly, neither of these differences are statistically significant at conventional levels. Furthermore, the net benefit statistic shows a positive net benefit at λ = £20 000, but is also

Table 5.12 Summary statistics for the cost-effectiveness of tight blood pressure control compared with less tight control in the UK prospective diabetes study

	Mean	SE	Lower 95% limit	Upper 95% limit
Control Group				
Effect (yr)	10.30	0.17	9.97	10.64
Cost (£)	6145	434	5294	6996
Treatment Group				
Effect (yr)	10.63	0.12	10.41	10.96
Cost (£)	6381	309	5775	6987
Difference				
Effect (yr)	0.33	0.21	−0.08	0.73
Cost (£)	236	533	−808	1280
ICER	720	N/A	N/A	N/A
INB*	6319	4169	−1853	14 490

*For λ = £20 000 per year of life gained.

non-significant. A naïve economic analysis might conclude that there was nothing to choose between these two treatments, however, this would ignore the wealth of information built up in the UK PDS study over an 11-year median follow-up time that the effectiveness of treatment is approaching the standard level of statistical significance. Indeed, the clinical endpoint of time to a diabetes-related event was shown to have a statistically significant difference between the two treatments. Instead the authors presented the results of their analysis in terms of a cost-effectiveness acceptability curve showing, for example, that for $\lambda = £20\,000/\text{LYG}$, there is a 92% probability that the intervention is cost-effective.

Of course, such an interpretation is only possible using a Bayesian view of probability. The quote from the original paper above indicates that the authors of the original study had considered that there was little information on the cost-effectiveness of tight blood pressure control for diabetic patients prior to the their reported results. However, a search of a database of cost-effectiveness analyses reporting cost per QALY and cost per life-year results (Briggs and Gray, 1999) for economic analyses of hypertension control conducted alongside a clinical trial identified an economic analysis of the Swedish Trial of Old patients with Hypertension (the STOP hypertension study; Dahlof *et al.*, 1991; Johannesson *et al.*, 1993). Clearly, the population studied (elderly Swedish patients without diabetes) differed from that of hypertensive patients in the UK PDS, but in the absence of other information, this trial might been seen as providing at least some information of the likely costs and effectiveness in the UK PDS population. Taking the average estimates of the life-years gained and costs reported in the economic analysis of the STOP study suggests that an estimated 0.16 life-years might be gained from treatment at an additional cost of approximately £1400 (converted from Swedish kronor and inflated to 1996 prices). Unfortunately, there was little information on sampling variation given in the economic analysis of the STOP study and standard errors on cost and effect differences were not given. Therefore, the standard errors were set arbitrarily to give a coefficient of variation on cost and effect equal to 2, in order to be conservative and to reflect the fact that the different methods employed in the economic analysis, the difference between the patient populations and the difference

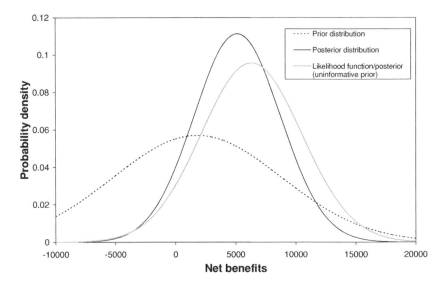

Figure 5.11 Bayesian approach to cost-effectiveness analysis of tight blood pressure control for hypertensive diabetic patients. Two posterior distributions for net benefits based on an informative prior distribution (shown) and an uninformative prior

between the two countries' health care systems will all increase the level of uncertainty associated with these prior estimates.

Using the net benefit approach, Figure 5.11 presents this prior information and the posterior distribution arising from employing this information together with the data from Table 5.12. This is the Bayesian approach with an informative prior. The alternative would be to employ an uninformative prior such that the posterior distribution produced is dominated by the observed data—either because no prior information is available, or because the analyst wishes to discard that information. The posterior distribution based on the uninformative prior is also shown in Figure 5.11, and exactly corresponds with the data likelihood. It is clear that incorporating the prior information reduces the variance of the posterior distribution, but that the point estimate of net benefit is weighted most heavily toward the UK PDS data.

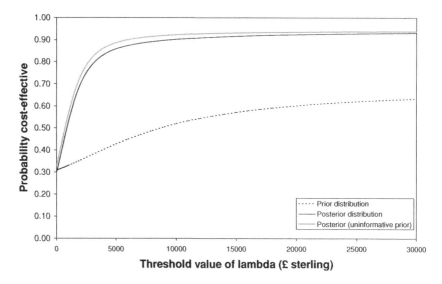

Figure 5.12 Bayesian approach to cost-effectiveness analysis of tight blood pressure control for hypertensive diabetic patients. Two posterior cost-effectiveness acceptability curves based on an informative prior distribution (shown) and an uninformative prior

Having estimated the prior and posterior distributions for net benefits using Bayesian methods, the probability of the intervention being cost-effective can then be plotted as a function of λ to generate cost-effectiveness acceptability curves and these are shown in Figure 5.12. Just as the Bayesian posterior distribution of net benefit with an uninformative prior corresponds to the data likelihood, so the cost-effectiveness acceptability curve plotted under the assumption of a uninformative prior will correspond exactly to a cost-effectiveness acceptability curve calculated by frequentist methods (involving the less natural interpretation of the curve based on *P*-values).

5.7   SUMMARY

This chapter has demonstrated different aspects of the material covered so far in the book using a range of different examples taken from real-life analyses. The aim was to illustrate and emphasize the importance of the methods discussed to this point. Having covered the basic methods for estimating the five key parameters required to summarize cost-effectiveness data, the remaining chapters move on to more specialized topics: power and sample size, covariate adjustment and subgroup analyses, multicenter studies, and statistical modeling of trial data.

6

Power and Sample Size Determination

6.1 INTRODUCTION

Issues regarding power and sample size determination for cost-effectiveness trials are discussed in this chapter. Two broad patterns of approach have been adopted in the literature, reflecting the analysis of ICERs and the CE plane versus the INB approach. The ICER/CE plane approaches are reviewed in Section 6.2, but in general provide a more cumbersome solution to the issue of power and sample size determination than does net benefit. Three approaches based on INB are outlined in the subsequent three sections. The first takes a classical approach based on the INB, limiting the type I and II errors. The second represents a Bayesian perspective on the classical sample size calculation based on posterior tail-area probabilities. The final approach is fully Bayesian and uses the expected value of information in an attempt to determine the sample size that maximizes the difference between the value of the information derived from a trial and the cost of conducting it.

Recall from Section 2.2 that C_{ji} is the total cost over the duration of interest for patient i on arm j, $j = T, S$. Let E_{ji} be the observed effectiveness for patient i on arm j. Then:

$$
\begin{aligned}
E_{ji} &= \bar{\delta}_{ji} && \text{for the probability of surviving (see Section 2.3.1)} \\
&= X_{ji} && \text{for mean survival (see Section 2.3.2)} \\
&= q_{ji} && \text{for mean quality-adjusted survival (see Section 2.3.3)}
\end{aligned}
$$

Statistical Analysis of Cost-effectiveness Data. A. Willan and A. Briggs
© 2006 John Wiley & Sons, Ltd.

Let

$$\Sigma_j = V\begin{pmatrix} E_{ji} \\ C_{ji} \end{pmatrix} = \begin{pmatrix} \sigma_j^2 & \rho_j\sigma_j\omega_j \\ \rho_j\sigma_j\omega_j & \omega_j^2 \end{pmatrix}$$

6.2 APPROACHES BASED ON THE COST-EFFECTIVENESS PLANE

Consider two threshold values for the WTP for a unit of effectiveness. The first, λ_L, is the largest value that would be considered a 'bargain'. That is, if the ICER was less than λ_L, and $\Delta_e > 0$, T would almost certainly be adopted. The second value, $\lambda_U > \lambda_L$, is the smallest value that would be considered prohibitive. That is, if the ICER was greater than λ_U, and $\Delta_e > 0$, S would almost certainly be retained. These threshold values are depicted in Figure 6.1.

In the shaded region R_0, T is considered not cost-effective. In the NW quadrant, T is not cost-effective regardless of the thresholds. For points in R_0 that lie in the NE quadrant, T is not cost-effective because the increase in effectiveness does not justify the increase in cost. Similarly, for points in R_0 that lie in the SW quadrant, T is not cost-effective because the decrease in effectiveness is not justified by the decrease in cost. In the shaded region R_1, T is considered cost-effective. In the SE quadrant, T is cost-effective regardless of the thresholds. For points in R_1 that lie in the NE quadrant, T is cost-effective because the increase in effectiveness justifies the increase in cost. Similarly, for points in R_1 that lie in the SW quadrant T is cost-effective because the decrease in effectiveness is justified by the decrease in cost.

If the truth lies in region R_0, we require a high probability, say $1 - \alpha$, of retaining S and if the truth lies in R_1, we require probability $1 - \beta$ of adopting, T. R_0 can be viewed as the null hypothesis and R_1 as the specific alternative hypothesis. Three main contributions on power and sample size determination have been made that pre-date the use of INB. The first was given by Briggs and Gray (1998b) using the confidence box approach of O'Brien *et al.* (1994). Another is given in Willan and O'Brien (1999) and the final approach follows methods given in Gardiner *et al.* (2000).

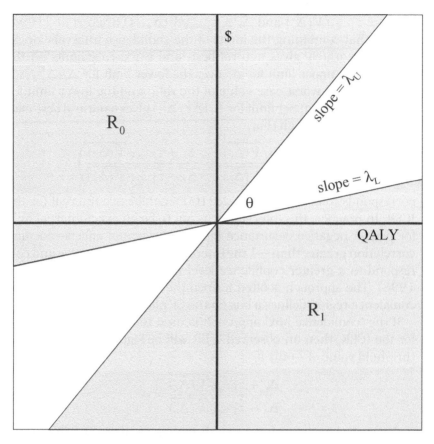

Figure 6.1 The cost-effectiveness plane with upper and lower thresholds

6.2.1   Briggs and Gray

The solution to the problem of sample size determination offered by Briggs and Gray (1998b) was based on an early solution to the problem of confidence interval calculation for the ICER offered by O'Brien *et al.* (1994). They proposed a simple (though conservative) solution based on a combination of the upper and lower limits on the standard confidence intervals for Δ_c and Δ_e. The $(1 - 2\alpha)100\%$ confidence limits employing standard parametric assumptions are given by $\hat{\Delta}_c \pm z_{1-\alpha}\sqrt{\hat{V}(\hat{\Delta}_c)}$ and $\hat{\Delta}_e \pm z_{1-\alpha}\sqrt{\hat{V}(\hat{\Delta}_e)}$. O'Brien *et al.* (1994) argued that combining the limits of the confidence intervals for Δ_c and Δ_e separately gives natural best- and worst-case limits on the ratio; i.e. the upper limit for Δ_c over the lower limit for $\Delta_e(\Delta_c^U/\Delta_e^L)$ gives a natural worst-case value of the ratio and the lower limit for Δ_c divided by the upper limit for $\Delta_e(\Delta_c^L/\Delta_e^U)$ gives a natural best-case value of the ratio, such that:

$$\left(\frac{\hat{\Delta}_c - z_{1-\alpha}\sqrt{\hat{V}(\hat{\Delta}_c)}}{\hat{\Delta}_e + z_{1-\alpha}\sqrt{\hat{V}(\hat{\Delta}_e)}}, \quad \frac{\hat{\Delta}_c + z_{1-\alpha}\sqrt{\hat{V}(\hat{\Delta}_c)}}{\hat{\Delta}_e - z_{1-\alpha}\sqrt{\hat{V}(\hat{\Delta}_e)}} \right)$$

corresponds to at least a $(1 - 2\alpha)100\%$ confidence interval for the ICER. In practice, this interval turns out to be an exact interval only for perfect negative covariance between costs and effects—for any correlation greater than -1 the interval will be conservative and correspond to a greater confidence level than 95% (Briggs and Fenn, 1998). The approach is often termed the 'confidence box' since the confidence region defines a box on the CE plane.

If the 'confidence box' approach is used to define an upper limit for the ICER, then an observed ICER will be significantly below the threshold value of λ only if

$$\frac{\Delta_c + z_{1-\alpha}\sqrt{\hat{V}(\hat{\Delta}_c)}}{\Delta_e - z_{1-\alpha}\sqrt{\hat{V}(\hat{\Delta}_e)}} < \lambda \qquad (6.2.1)$$

At the design stage of an analysis it is necessary to specify the hypothesized difference in cost, δ_c and the hypothesized difference in effectiveness δ_e that it is hoped to detect (at the required significance level). Clearly, at this stage, it must be the case that $\delta_c/\delta_e < \lambda$, otherwise any trial to show cost-effectiveness would be considered futile. This emphasizes that, while the analyst can avoid explicit identification of the threshold value of λ at the analysis stage, at the design stage they cannot, and any sample size formula must specify λ in addition to other parameters.

Following a similar logic to the case of sample size calculations for a single outcome (either cost or effectiveness) Briggs and Gray go on to argue that to have a power of $1 - \beta$ for achieving significance at the $2\alpha 100\%$ level as in Equation (6.2.1), the following inequality must

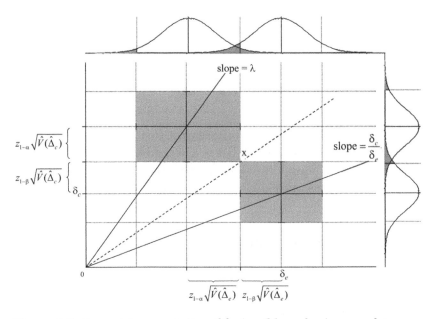

Figure 6.2 Geometric presentation of the 'confidence box' approach to sample size calculations on the CE plane

be satisfied

$$\frac{\delta_c + z_{1-\alpha}\sqrt{\hat{V}(\hat{\Delta}_c)} + z_{1-\beta}\sqrt{\hat{V}(\hat{\Delta}_c)}}{\delta_e - z_{1-\alpha}\sqrt{\hat{V}(\hat{\Delta}_e)} - z_{1-\beta}\sqrt{\hat{V}(\hat{\Delta}_e)}} < \lambda \qquad (6.2.2)$$

They show how this inequality can be given a geometric interpretation on the CE plane, and this is shown in Figure 6.2. For an observed ICER to be significantly below a given a threshold value of λ, it must lie below and to the right of the point x in Figure 6.1. For the power to be greater than $1 - \beta$, the point x in Figure 6.1 must lie above and to the left of the $(1 - 2\beta)$ 100% confidence region. The distributions under the null and alternative hypotheses are shown at the top of the plane for $\hat{\Delta}_e$ and to the right of the plane for $\hat{\Delta}_c$ in order to clarify the construction of Equation (6.2.2). Substituting in the standard errors, assuming an equal number of patients per arm, yields the sample size formula

$$n > \left[\frac{(z_{1-\alpha} + z_{1-\beta}) \left(\lambda \sqrt{\sigma_T^2 + \sigma_S^2} + \sqrt{\omega_T^2 + \omega_S^2} \right)}{\delta_c - \lambda \delta_e} \right]^2$$

Of course, this formula is conservative, just as the confidence box approach method is. Also, it bears a resemblance to the net benefit formulation (Equation 6.3.1) given in Section 6.3.1, and the two formulae are the same if $\sigma_T^2 / \sigma_S^2 = \omega_T^2 / \omega_S^2$ and $\rho_T = \rho_S = -1$.

6.2.2   Willan and O'Brien

In this approach we set $\beta = \alpha$ and require that the angle between the upper and lower limits of the $100(1 - \alpha)\%$ confidence interval for the ICER must be no greater than θ, as shown in Figure 6.1. To achieve this, the sample size required in each arm is given by,

$$n = \left[\left\{ \left[z_{1-\alpha/2} \Big/ \sin \left\{ \tfrac{1}{2} \cos^{-1} \left(m_L^T G^{-1} m_U \right. \right. \right. \right. \\ \left. \left. \left. \left. \Big/ \sqrt{m_L^T G^{-1} m_L m_U^T G^{-1} m_U} \right) \right\} \right]^2 - X_1^2 \right\}^{1/2} - Z \right]^2 \Big/ \Delta^T G^{-1} \Delta$$

where $m_k = (1, \lambda_k)^T$, $G = \Sigma_T + \Sigma_S$, X_1^2 is a chi-squared random variable with one degree of freedom, Z is a standard normal random variable and $\Delta = (\Delta_e, \Delta_c)^T$ (see Willan and O'Brien, 1999). Unhelpfully, the formula for the required sample size contains the true values of Δ_e and Δ_c and two random variables. This is because the angle between the upper and lower limits of the confidence interval will depend on where the observed values $\hat{\Delta}_e$ and $\hat{\Delta}_c$ fall on the CE-plane. (The closer to the origin that $\hat{\Delta}_e$ and $\hat{\Delta}_c$ are, the wider the angle.) Therefore, some adjustments must be made.

As the values of X_1^2 and Z increase, the sample size increases. Therefore, adopting a conservative approach, for some $\gamma > 0.5$, Z is replaced by z_γ and X_1^2 is replaced by $x_{1,\gamma}^2$ where $x_{1,\gamma}^2$ is the 100γth percentile for a chi-square distribution with one degree of freedom. Since X_1^2 and Z are independent, the probability that the sample size is sufficiently large is γ^2. Rearranging the equation yields

$$\Delta^T G^{-1} \Delta = \left[\left\{ \left[z_{1-\alpha/2} \Big/ \sin \left\{ \tfrac{1}{2} \cos^{-1} \left(m_L^T G^{-1} m_U \right. \right. \right. \right. \right.$$
$$\left. \left. \left. \left. \Big/ \sqrt{m_L^T G^{-1} m_L m_U^T G^{-1} m_U} \right) \right\} \right]^2 - x_{1,\gamma}^2 \right\}^{1/2} - z_\gamma \right]^2 \Big/ n$$

For a given sample size of n per arm the above equation defines an ellipse on the CE plane centered at the origin. If the truth lies outside the ellipse, then the sample size is adequate. However, if the truth lies inside the ellipse, it is not. Therefore, for a given n, investigators can determine the region on the CE plane for which n is adequate. If within the ellipse there are important differences in cost or effectiveness, regardless of the resulting ICER, then a larger sample size is required. It is important to note that, although this method is based on providing a sufficiently narrow confidence interval for the ICER, the adequacy of a particular sample size is based on having adequate power for minimal differences in costs and effectiveness.

Another problem is the availability of estimates of Σ_T and Σ_S. If there are few previous data for patients receiving T, the assumption that $\Sigma_T = \Sigma_S$ can be made, i.e. $G = 2\Sigma_S$. If there are few cost data for patients receiving S, then ρ_S could be set to zero and a retrospective collection of key cost determinates, such as days in hospital and number of major procedures, on a sample of patients could be used to get a rough estimate of ω_S^2. In any event, updated estimates of Σ_T and Σ_S can be made during the progress of the trial.

6.2.2.1   Example

Using the data from the trial of symptomatic, hormone-resistant prostate cancer introduced in Section 5.3, we have

$$\Sigma_T = \begin{pmatrix} 1470 & 169058 \\ 169058 & 394447420 \end{pmatrix}$$

$$\Sigma_S = \begin{pmatrix} 870.1 & 152444 \\ 152444 & 417252112 \end{pmatrix} \quad \text{and}$$

$$G = \begin{pmatrix} 2340 & 321502 \\ 321502 & 811699532 \end{pmatrix}$$

In Canadian dollars let $\lambda_L = 400/\text{QALW}$ and $\lambda_U = 2000/\text{QALW}$. These values approximately correspond to $2000/\text{QALY}$ and $100\,000/\text{QALY}$, respectively, which are thresholds proposed by Laupacis *et al.* (1992). Although these values may be somewhat out of date, they will suffice for illustration. Setting $\alpha = 0.05$ and $\gamma = 0.9$, the ellipse is defined by the equation

$$(\Delta_e, \Delta_c) \begin{pmatrix} 2340 & 321502 \\ 321502 & 811699532 \end{pmatrix}^{-1} \begin{pmatrix} \Delta_e \\ \Delta_c \end{pmatrix} = 5.707648/n$$

which simplifies to

$$451.9\Delta_e^2 - 0.358\Delta_e\Delta_c + 0.0013029\Delta_c^2 - 5707\,648/n = 0.$$

The 'power' ellipses for $n = 100$, 250 and 1000 are illustrated in Figure 6.3. For reference the observed (Δ_e, Δ_c) is plotted. By inspection one can see that, for a sample size of 100 per arm, there are sizable differences in cost and effectiveness for which power is inadequate to some extent this is also true for $n = 250$. Perhaps only for $n = 1000$

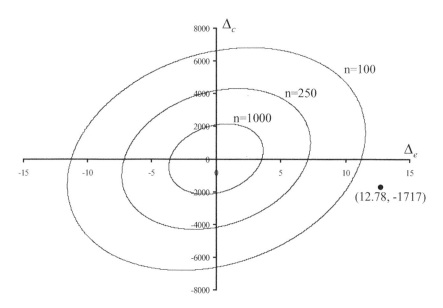

Figure 6.3 Power ellipses for prostate example

is the power adequate, with the maximum differences for which the sample size is inadequate, being CAD$2000 and 4 QALW for cost and effectiveness, respectively.

6.2.3 Gardiner *et al.*

Referring again to Figure 6.1, let R_0^+ and R_1^+ be that part of the region R_0 and R_1, respectively, for which $\Delta_e > 0$. Therefore, R_0^+ is the region defined by $\Delta_c/\Delta_e > \lambda_U$ and $\Delta_e > 0$, and R_1^+ is the region defined by $\Delta_c/\Delta_e < \lambda_L$ and $\Delta_e > 0$. Assuming $\Delta_e > 0$, and following along the lines of Gardiner *et al.* (2000), consider testing the hypothesis R_0^+ versus R_1^+, by rejecting R_0^+ if $\hat{\Delta}_e\lambda_U - \hat{\Delta}_c > k^+ > 0$. That is, reject R_0^+ in favor of R_1^+ if the observed INB for a willingness-to-pay of λ_U is sufficiently positive. We require that

$$\Pr(\hat{\Delta}_e\lambda_U - \hat{\Delta}_c > k^+ | R_0^+) \leq \alpha \qquad \text{and}$$
$$\Pr(\hat{\Delta}_e\lambda_U - \hat{\Delta}_c > k^+ | R_1^+) \geq 1 - \beta$$

Or, equivalently, that

$$\Pr((\hat{\Delta}_e\lambda_U - \hat{\Delta}_c)/\Delta_e > k^+/\Delta_e | R_0^+) \leq \alpha \qquad \text{and}$$
$$\Pr((\hat{\Delta}_e\lambda_U - \hat{\Delta}_c)/\Delta_e > k^+/\Delta_e | R_1^+) \geq 1 - \beta$$

Given that n^+ patients are randomized to each arm, and assuming normality, the least favorable distributions for $(\hat{\Delta}_e\lambda_U - \hat{\Delta}_c)/\Delta_e$ under R_0^+ and R_1^+, respectively, are given by

$$N\left(0, \frac{1}{\Delta_e^2}\frac{\sigma^2(\lambda_U)}{n^+}\right) \qquad \text{and} \qquad N\left(\lambda_U - \lambda_L, \frac{1}{\Delta_e^2}\frac{\sigma^2(\lambda_L)}{n^+}\right)$$

where $\sigma^2(\lambda) = (\sigma_T^2 + \sigma_S^2)\lambda^2 + \omega_T^2 + \omega_S^2 - 2(\rho_T\sigma_T\omega_T + \rho_S\sigma_S\omega_S)\lambda$. To meet the requirements, the upper $100(1 - \alpha)$th percentile for the distribution for $(\hat{\Delta}_e\lambda_U - \hat{\Delta}_c)/\Delta_e$ under R_0^+ must equal the 100βth percentile for the distribution under R_1^+. Therefore

$$\frac{z_{1-\alpha}\sigma(\lambda_U)}{\Delta_e\sqrt{n^+}} = (\lambda_U - \lambda_L) - \frac{z_{1-\beta}\sigma(\lambda_L)}{\Delta_e\sqrt{n^+}}$$

yielding

$$n^+ = \left\{ \frac{z_{1-\alpha}\sigma(\lambda_U) + z_{1-\beta}\sigma(\lambda_L)}{\Delta_e\,(\lambda_U - \lambda_L)} \right\}^2$$

as the required sample size when $\Delta_e > 0$.

For the other side of the CE-plane, let R_0^- and R_1^- be that part of the region R_0 and R_1, respectively, for which $\Delta_e < 0$. Therefore, R_0^- is the region defined by $\Delta_c/\Delta_e < \lambda_L$ and $\Delta_e < 0$, and R_1^- is the region defined by $\Delta_c/\Delta_e > \lambda_U$ and $\Delta_e < 0$. Assuming $\Delta_e < 0$, consider testing the hypothesis R_0^- versus R_1^-, by rejecting R_0^- if $\hat{\Delta}_e\lambda_L - \hat{\Delta}_c > k^- > 0$. That is, reject R_0^- in favor of R_1^- if the observed INB for a willingness-to-pay of λ_L is sufficiently positive. We require that

$$\Pr(\hat{\Delta}_e\lambda_L - \hat{\Delta}_c > k^- | R_0^-) \le \alpha \qquad \text{and}$$
$$\Pr(\hat{\Delta}_e\lambda_L - \hat{\Delta}_c > k^- | R_1^-) \ge 1 - \beta$$

This leads to a required sample size when $\Delta_e < 0$ of

$$n^- = \left\{ \frac{z_{1-\alpha}\sigma(\lambda_L) + z_{1-\beta}\sigma(\lambda_U)}{\Delta_e\,(\lambda_U - \lambda_L)} \right\}^2$$

Therefore the required sample size depends on whether Δ_e is positive or negative, and since this is unknown, the required sample size should be set to the maximum of n^+ and n^-. However, if we set $\beta = \alpha$, then

$$n^+ = n^- = \left\{ \frac{z_{1-\alpha}\left[\sigma(\lambda_U) + \sigma(\lambda_L)\right]}{\Delta_e\,(\lambda_U - \lambda_L)} \right\}^2$$

Setting $\beta = \alpha$ does not solve the issue that the sample size required depends on the true value of Δ_e. As Δ_e approaches 0, the required sample size becomes very large, which reflects the fact that inference about the ICER is problematic when $\hat{\Delta}_e$ is close to zero in probability. Therefore, the best we can do is to determine a sample size that is adequate if the true difference in effectiveness is greater, in absolute terms, than some pre-specified amount, perhaps the smallest clinically important difference, and recognize that there is insufficient power if the difference is less. If investigators expect Δ_e to be close to zero, as might be expected in a cost-minimization trial, sample size determinations based on the INB, as shown in Section 6.2, are more appropriate.

6.2.3.1   Example

Using the data from the trial of symptomatic, hormone-resistant prostate cancer introduced in Section 5.3 and discussed in Section 6.2.2.1, we have

$$\sigma(400) = \sqrt{400^2 \times 2340 + 811\,699\,532 - 2 \times 400 \times 321\,502}$$
$$= 30\,478$$

and

$$\sigma^2(2000)$$
$$= \sqrt{2000^2 \times 2340 + 811\,699\,532 - 2 \times 2000 \times 321\,502}$$
$$= 94\,264.$$

Setting $\alpha = \beta = 0.05$, the required samples size is given by

$$n = \left\{ \frac{1.65\,(94\,264 + 30\,478)}{\Delta_e\,(2000 - 400)} \right\}^2 = \frac{16\,548}{\Delta_e^2}$$

Therefore, if the difference in effectiveness is at least 5, then the required sample size is 662 patients per arm.

Rearranging the equation to read $\Delta_e = \sqrt{16548/n}$ and substituting for n the values 100, 250 and 1000, yield values for Δ_e of 12.9, 8.14 and 4.07, respectively. These values correspond closely to the horizontal intercepts of the ellipses in Figure 6.2, derived from the Willan and O'Brien approach.

6.3   THE CLASSICAL APPROACH BASED ON NET BENEFIT

6.3.1   The method

In an effectiveness (i.e. management or pragmatic) trial patients are randomized between S and T with the purpose of adopting T if the (one-sided) null hypothesis is rejected. The classical sample size determinations are based on achieving a power curve with predetermined specifications. The power curve, which is the probability of rejecting the null hypothesis (i.e. adopting T) as a function of the true treatment difference, is required to pass through the null hypothesis at a specified value, often denoted as α and referred to as the type I error. Thus, the type I error is the probability of adopting T if the null hypothesis is true. To achieve this specification, the level (size) of the test of the null hypothesis must be set to the type I error. This forces the power curve to pass through the type I error at the null hypothesis, regardless of the sample size. The sample size is set by specifying that the power curve pass through $1 -$ type II error at some predetermined value which represents the smallest clinically important difference, often denoted as δ. The type II error is the probability of failing to reject the null hypothesis, and adopting T, when the true treatment difference is equal to δ. The probability of the type II error is denoted as β, and $1 - \beta$ is referred to as the power of the test of the null hypothesis. The power is the lower bound of the probability of adopting T if the true treatment difference is clinically important. Thus, by specifying α, β and δ and having some estimate of the between-patient variation of the primary outcome, the sample size can be determined (see Lemeshow *et al.*, 1990 or Lachin, 1981). The between-patient variance of net benefit for arm j is given by $V(\lambda E_{ji} - C_{ji}) = \sigma_j^2 \lambda^2 + \omega_j^2 - 2\rho_j \sigma_j \omega_j \lambda \equiv v_j$. Therefore the sample size per arm required to have a $100(1 - \beta)\%$ probability of rejecting the null hypothesis of no difference at the one-sided α level, if INB equals δ, is given by

$$n = (z_{1-\alpha} + z_{1-\beta})^2 (v_T + v_S)/\delta^2 \tag{6.3.1}$$

The power curve, which plots the probability of rejecting the null hypothesis as a function of δ, for a fixed sample size, is illustrated in Figure 6.4. Inspection of Figure 6.4 reveals that, for values of INB between 0 and $\delta_{0.5}$, there is less than 50% probability of rejecting the null hypothesis, and presumably adopting T, even if the true INB is positive. Because of this and because treatment arms are being compared with respect to *net* benefit, some authors have recommended setting $\alpha = 0.5$ (see Claxton, 1999 and Willan, 1994). In which case, $z_{0.5} = 0$ and the sample size formula becomes $n = z_{1-\beta}^2 (v_T + v_S)/\delta^2$, affording a substantial reduction. Analysis is simplified in this approach and T is adopted if the observed INB ($\hat{\Delta}_e \lambda - \hat{\Delta}_c$) is positive. This guarantees that if the true INB is positive, there is at least a 50%

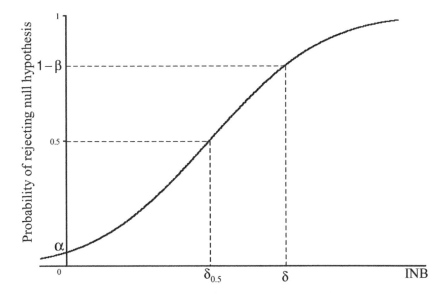

Figure 6.4 Traditional power curve

probability of adopting T, and if the true INB is zero or negative, there is at least a 50% probability of retaining S. A fuller discussion of this approach can be found in Section 6.5.

The choice of δ is often difficult. In theory it is the smallest difference in favor of T for which the investigators would like to have a high probability of rejecting the null hypothesis. Because treatment arms are being compared with respect to *net* benefit on a universal monetary scale, the argument could be made that δ should be the same for all trials in all therapeutic areas. After all, a $1000 increase in net benefit for a patient being treated for congestive heart failure has the same value as a $1000 increase in net benefit for a patient being treated for cancer.

6.3.2 Example

Using the data from the trial of symptomatic, hormone-resistant prostate cancer introduced in Section 5.3 and discussed in Sections

Table 6.1   Sample size per arm for INB approach for prostate example

Level (α)	Willingness-to-pay (λ)			
	400	1000	1500	2000
0.05	2555	6899	14 056	24 428
0.5	293	790	1610	2799

6.2.2.1 and 6.2.3.1, the between-patient variances of INB are

$$v_T = \lambda^2 1470 + 394447420 - 338116\lambda \qquad \text{and}$$
$$v_S = \lambda^2 870.1 + 417252112 - 304888\lambda$$

Setting $\beta = 0.2$ and $\delta = 1500$, the required sample size per arm for various values of λ is given in Table 6.1 for $\alpha = 0.05$ and 0.5. Sample size requirements increase with λ, since the variance does. And, of course, setting $\alpha = 0.5$ reduces sample size requirements substantially.

6.4   BAYESIAN TAKE ON THE CLASSICAL APPROACH

6.4.1   The Method

A Bayesian assessment of sample size determination for cost-effectiveness trials is given in O'Hagan and Stevens (2001). They state two objectives. The first, referred to as the *analysis objective*, is that the outcome of the trial will be considered positive if the posterior probability that INB is positive (i.e. T is cost-effective) is at least $1 - \alpha$. The second, referred to as the *design objective*, is that the sample size should be sufficiently large that the probability of achieving a positive result is at least $1 - \beta$. They propose a novel Bayesian solution based on assuming two prior distributions: one for the design stage and one for the analysis stage.

Simplifying the notation by dropping the subscript λ, let $b = \lambda \Delta_e - \Delta_c$ and $v = (\sigma_T^2 + \sigma_S^2) \lambda^2 + \omega_T^2 + \omega_S^2 - 2 (\rho_T \sigma_T \omega_T + \rho_S \sigma_S \omega_S) \lambda$ (note that $v = \sigma^2(\lambda)$ from Section 6.2.3 and $v_T + v_S$ from Section 6.3.1). Further, let the prior mean and variance at the design stage be b_d and v_d and at the analysis stage be b_a and v_a. O'Hagan and Stevens show that the analysis objective can be met by declaring T cost-effective if, and only if,

$$\sqrt{v^*} \left(v_a^{-1} b_a + n v^{-1} \hat{b} \right) \geq z_{1-\alpha} \qquad (6.4.1)$$

where $\hat{b}$ is the estimate of b from the trial, $v^* = \left(v_a^{-1} + n v^{-1} \right)^{-1}$ and n is the sample size per arm. They also show that to meet the design objective n must be large enough to satisfy the inequality

$$\left(v_a^{-1} b_a + n v^{-1} b_d \right) - \frac{z_{1-\alpha}}{\sqrt{v^*}} - z_{1-\beta} \sqrt{\left(n v^{-1} \right)^2 \left(v_d + n^{-1} v \right)} \geq 0$$
$$(6.4.2)$$

Inequality (6.4.2) has a finite solution if, and only if, $\Phi(b_d/\sqrt{v_d}) > 1 - \beta$, where $\Phi(\cdot)$ is the cumulative probability distribution for the standard normal distribution. That is to say, that no sample size is large enough to guarantee a positive result with $1 - \beta$ probability, if the prior probability that T is cost-effective at the design stage ($\Phi(b_d/\sqrt{v_d})$) is less than $1 - \beta$.

For the situation where $v_a^{-1} = 0$ (i.e. analysis prior is *weak*) and $v_d = 0$ (i.e. design prior is *strong*), we have the frequentist solution, and Inequality (6.4.1) becomes $\hat{b} \geq z_{1-\alpha} \sqrt{v/n}$ and the solution to Inequality (6.4.2) is $n = (z_{1-\alpha} + z_{1-\alpha})^2 v/b_d^2$.

6.4.2 Example

Using the data from the prostate trial, introduced in Section 5.3 and discussed in Sections 6.2.2.1, 6.2.3.1 and 6.3.2, for $\lambda = 400$, we have $v = 400^2 \times 2340 + 811\,699\,532 - 2 \times 400 \times 321\,502 = 30\,478^2$. Also, we let $b_a = 12.78 \times 400 - (-1717) = 6829$ and $v_a = 400^2 \times 40.52 + 14\,339\,032 - 2 \times 400 \times 5647 = 4038^2$, which are the posterior mean and variance from the existing trial. Setting $\alpha = 0.05$ and $\beta = 0.2$ and assuming a strong prior at the design stage, say $b_d = 1500$ and $v_d = 0$, Inequality (6.4.2) reduces to

$$\left(v_a^{-1}b_a + nv^{-1}b_d\right) - \frac{z_{1-\alpha}}{\sqrt{v^*}} - z_{1-\beta}\sqrt{nv^{-1}} \geq 0$$

yielding

$$-\frac{\left(4038^{-2} \times 6829 + n30478^{-2} \times 1500\right)}{\sqrt{\left(4038^{-2} + n30478^{-2}\right)^{-1}}} - 0.8416\sqrt{n30478^{-2}} \geq 0$$

The smallest n satisfying this inequality is 2050.

The number of patients required per arm using the frequentist approach (i.e. with $v_a^{-1} = 0$) with the same α and β and smallest clinically important difference δ of 1500 is given in Table 6.1 as 2555. The reduction of sample size using the Bayesian approach reflects the intention, at the analysis stage, of using the informative prior derived from the existing trial.

6.5   THE VALUE OF INFORMATION APPROACH

6.5.1   The method

Although the classical approach taken in Section 6.3 is logical and has intuitive appeal, the choices of α, β and δ are arbitrary at best, and absurd at worse. The value of α is most often set to 0.05, regardless of the cost of making a type I error. Consequently, a trial that randomizes patients with age-related macular degeneration between two different wavelengths of laser coagulation (see Willan *et al.* 1996), has the same probability of falsely declaring T superior as does a trial of cesarean section versus vaginal delivery for women presenting in the breech position (see Hannah *et al.*, 2000). In the former there is little or no cost to declaring one wavelength superior to another when they are equally effective. However, in the latter, declaring cesarean section superior, when it is the same as vaginal delivery, carries substantial cost. Assigning the same probability to the two errors is absurd, quite apart from the fact that the value of 0.05 is arbitrary in the first place. Also arbitrary is the typical choice of 0.2 for the type II error. It means that there is a 20% chance that the effort and money invested in the trial will be wasted, even if there is a clinically important difference between the treatments. And again, it fails to reflect the cost of making the error. The choice of δ can be less arbitrary and can be estimated by polling clinicians and decision makers. However, in practice it is often back-solved from the sample size equation after substituting in a sample size that primarily reflects feasibility and cost considerations. Even if δ is a reasonable estimate of the smallest clinically important difference, there is a range of values for the true treatment difference that is less than the smallest clinically important difference for which the chance of rejecting the null hypothesis, and adopting T, is greater than 50%. This range is shown in Figure 6.4 as lying between $\delta_{0.5}$ and δ.

In some sense the point $(\delta_{0.5}, 0.5)$ can be recognized as the only non-arbitrary point on the power curve. The value $\delta_{0.5}$ should represent the difference for which clinicians and decision makers are indifferent between treatments, since for true differences greater than $\delta_{0.5}$ there is at least a 50% chance that the null hypothesis will be rejected and T adopted, and for true differences less than $\delta_{0.5}$ the chance that the null hypothesis will not be rejected and S retained is less than 50% (see Willan, 1994).

In response to this arbitrariness many authors have proposed alternative approaches. In particular, see Claxton and Posnett (1996) and Willan and Pinto (2005). A review is provided by Pezeshk (2003). Most authors have proposed using a Bayesian decision theoretic approach, introduced by Grundy *et al.* (1956) and developed by Raiffa and Schlaifer (2000), to determine the sample size that maximizes the difference between the cost of the trial and the expected value of the results. The expected value of the results from the trial is measured by the expected value of sample information, and the costs considered are the financial cost and the expected opportunity loss of treating patients in the trial with the inferior treatment. The perspective taken is that of a society that pays for health care through a single payer system and also pays for the financial cost of the trial through government or private donation-based or philanthropic agencies. More detail is given in Willan and Pinto (2005).

Taking a Bayesian approach, assume that the prior distribution for INB is normal with mean b_0 and variance v_0, and letting $f_0(\cdot)$ be the prior density function. The prior information is most likely to be a synthesis of one or more previous relevant trials. If no other data are to be sought, the decision rule that minimizes opportunity loss is: adopt T if the b_0 is positive and retain S otherwise. This rule also forces the power curve through the point $(0, 0.5)$ and eliminates the need for statistical inference (see Claxton, 1999 and Willan, 1994).

The opportunity loss for a particular decision is the utility (in this case, net benefit) for the best decision minus the utility for the decision taken. Therefore, if b is positive, the opportunity loss per patient of adopting T is equal to 0, since adopting T is the best decision when b is positive. On the other hand, if b is negative, the opportunity loss per patient of adopting T is equal to $-b$, since the best decision is to retain S, and the utility (i.e. net benefit) of retaining S minus the utility of adopting T is $-b$ by definition. By similar arguments the opportunity loss per patient of retaining S is 0 if b is negative and b if b is positive. Here we dropped the subscript λ from b_λ for the sake of simplicity. The loss functions are illustrated in Figure 6.5. If there are N future patients for whom the decision is to be made then total expected opportunity loss is

$$EVPI_0 = N \left\{ I\{b_0 > 0\} \int_{-\infty}^{0} -b \, f_0(b) \, db + I\{b_0 \leq 0\} \int_{0}^{\infty} b \, f_0(b) \, db \right\}$$

$$= N\mathcal{D}(b_0, v_0) \tag{6.5.1}$$

where $\mathcal{D}(b_0, v_0) = [v_0/(2\pi)]^{\frac{1}{2}} \exp[-b_0^2/(2v_0)] - b_0[\Phi(-b_0/v_0^{\frac{1}{2}}) - I\{b_0 \leq 0\}]$ and $\Phi(\cdot)$ is the cumulative distribution function for a standard normal random variable. The term $\int_{-\infty}^{0} -b \, f_0(b) \, db$ is the expected opportunity loss for a single patient if T is adopted and $\int_{0}^{\infty} b \, f_0(b) \, db$ is the expected opportunity loss for a single patient if S is retained. $EVPI_0$ is referred to as the expected value of perfect information (EVPI), because the loss could be avoided in the presence

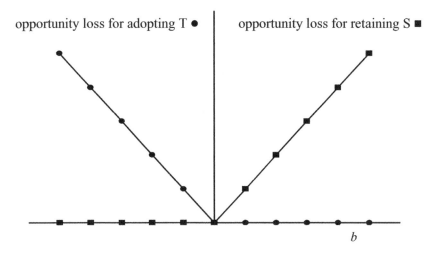

opportunity loss for adopting T ●   opportunity loss for retaining S ■

b

Figure 6.5 Opportunity loss functions

of perfect information about b. EVPI$_0$ is the maximum it is worth paying for more information about b (i.e. another trial), otherwise the cost of doing another trial would be greater than the expected loss incurred by making a decision based on current knowledge.

Suppose a trial of size n per arm is proposed, and let $\hat{b}$ be the estimator of b using the data from the new trial. Assuming normality, $\hat{b} \sim n\left(b_0, \ v_0 + 2\sigma^2/n\right)$, with density function $\hat{f}(\cdot)$, where σ^2 is the between-patient variance of net benefit, which is assumed known and the same in both treatment arms (i.e. $\sigma^2 = \sigma_T^2 \lambda^2 + \omega_T^2 - 2\rho_T \sigma_T \omega_T \lambda = \sigma_S^2 \lambda^2 + \omega_S^2 - 2\rho_S \sigma_S \omega_S \lambda$). Therefore, the posterior distribution for b is normal with mean

$$b_1 = v_1 \left(\frac{b_0}{v_0} + \frac{n\hat{b}}{2\sigma^2} \right)$$

and variance

$$v_1 = \left(\frac{1}{v_0} + \frac{n}{2\sigma^2} \right)^{-1}$$

Let the posterior density function be $f_1(\cdot)$. The post-trial EVPI is

$$EVPI_1 = (N - 2n) \left\{ I(b_1 > 0) \int_{-\infty}^{0} -b \, f_1(b) \, db + I(b_1 \le 0) \int_{0}^{\infty} b \, f_1(b) \, db \right\} = (N - 2n) \, \mathcal{D}(b_1, v_1)$$

The reduction in the expected value of perfect information, $EVPI_0 - EVPI_1$, is referred to as the expected value of sample information (EVSI) and is simply the value of the trial, since it is the amount by which the expected opportunity loss of making a decision is reduced. EVSI is a function of n and $\hat{b}$, and a trial that costs more than

$$E_{\hat{b}} \, EVSI(n, \hat{b}) = \int_{-\infty}^{\infty} EVSI(n, \hat{b}) \, \hat{f}(\hat{b}) \, d\hat{b}$$

$$= (N - 2n) \left\{ \mathcal{D}(b_0, v_0) - \int_{-\infty}^{\infty} \mathcal{D}(b_1, v_1) \, \hat{f}(\hat{b}) \, d\hat{b} \right\}$$

should not be performed. The optimal sample size is that which maximizes the difference between the $E_{\hat{b}} \, EVSI(n, \hat{b})$ and the cost of performing the trial.

An analytic solution for $\int_{-\infty}^{\infty} \mathcal{D}(b_1, v_1) \hat{f}(\hat{b}) d\hat{b}$ is given by $I_1 + I_2 + I_3$, where

$$I_1 = \sqrt{\frac{2v_0}{\pi}} \frac{\sigma^2 e^{\left(-\frac{b_0^2}{2v_0}\right)}}{n\sigma_{\hat{b}}^2} \qquad I_2 = -b_0 \Phi\left(-\frac{b_0}{\sqrt{v_0}}\right) + \frac{v_0^{3/2} e^{\left(-\frac{b_0^2}{2v_0}\right)}}{\sigma_{\hat{b}}^2 \sqrt{2\pi}}$$

$$I_3 = b_0 \Phi\left(-\frac{b_0\sqrt{\sigma_{\hat{b}}^2}}{v_0}\right) - \frac{v_0 e^{\left(-\frac{b_0^2\sigma_{\hat{b}}^2}{2v_0^2}\right)}}{\sqrt{2\pi\sigma_{\hat{b}}^2}}$$

and $\sigma_{\hat{b}}^2 = V(\hat{b}) = v_0 + 2\sigma^2/n$.

The cost of a trial has two components, one financial and the other reflecting opportunity loss. Let C_f be the fixed financial cost of setting up a trial and let C_v be the financial cost per patient. The total cost of the trials is

$$TC(n) = C_f + 2nC_v + n\,|b_0|$$

The term $n\,|b_0|$ is the total cost corresponding to the opportunity loss for the patients randomized to the arm which is currently observed to be inferior (i.e. S if $b_0 > 0$ and T is $b_0 \le 0$).

The expected net gain from the trial is $\text{ENG}(n) \equiv E_b\,\text{EVSI}(n,\hat{b}) - TC(n)$. For consistency, $TC(0)$ is set to 0, i.e. no fixed costs are incurred if no trial is done. The optimal sample size is that non-negative value of n that maximizes $\text{ENG}(n)$. The situation is illustrated in Figure 6.6. $TC(n)$ is a straight line with slope $2C_v + |b_0|$. $E_b\text{EVSI}(n,\hat{b})$ passes through the points $(0, 0)$ and $(N/2, 0)$ and has a single maximum. The optimal n is at the point where the slopes are equal, and is shown in Figure 6.6 as n^*. The vertical distance between $E_b\,\text{EVSI}(n^*, \hat{b})$ and

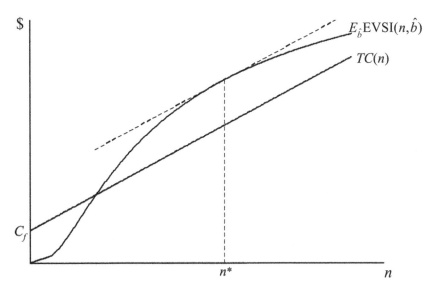

Figure 6.6 Total cost and the expected value of sample information as a function of sample size

$TC(n^*)$ is the optimum $ENG(n^*)$. If $TC(n) > E_{\hat{b}} EVSI(n, \hat{b})$ for all positive n, then the optimal sample size is 0. When the optimal sample size is 0 the current information (b_0 and v_0) are sufficient for decision making, and one could claim that we are not in a state of equipoise. On the other hand if the optimal sample size is positive, one could claim to be in a state of equipoise, since the expected value of the information from the trial, which takes into account the value placed on a unit of effectiveness, the current information and the number of future patients that would benefit, exceeds the cost of getting it.

Extensions of these methods that allow for unequal sample sizes and discounting of future costs and benefits can be found in Willan and Pinto (2005).

6.5.2 Example

In a pilot study by Hutton *et al.* (2003), 232 pregnant women presenting in the breech position were randomized between early (T) versus late (S) external cephalic version (ECV). ECV is an attempt to manipulate the fetus to a cephalic presentation. Elective caesarean section is accepted practice for breech presentation, and the primary outcome for the trial was a non-caesarean delivery. In the early ECV arm 41 of 116 patients (35.3%) had a non-caesarean delivery and in the late ECV arm the corresponding numbers are 33 of 116 (28.4%). Based on these data the investigators designed a larger trial of 730 patients per arm to have an 80% probability of rejecting the null hypothesis of no treatment effect, if the treatments differed by 8 percentage points, using a two-sided type I error of 0.05.

Suppose, for sake of argument, that society is willing to pay $1000 to achieve a non-caesarean delivery in these patients. This figure reflects the cost savings and the preference for a non-caesarean birth. Suppose further that, apart from the possible cost savings from preventing a caesarean delivery, there is no difference in cost between early and late ECV. Therefore, $b = \Delta_e 1000$, where Δ_e is the probability of a non-caesarean delivery for early ECV minus the probability of a non-caesarean delivery for late ECV. The prior distribution for b, given the pilot data, is assumed normal with mean

$$b_0 = (41/116 - 33/116)1000 = 68.97$$

and variance

$$v_0 = \left\{ \frac{41/116(1 - 41/116)}{116} + \frac{33/116(1 - 33/116)}{116} \right\} 1000^2$$
$$= 3724.78.$$

Using an overall non-caesarean delivery of $(41 + 33)/(116 + 116)$ $= 74/232$, an estimate of the between patient variance is

$$\sigma^2 = 74/232(1 - 74/232)1000^2 = 217\,227$$

Assuming that 50% of the North American patients in the 20-year period following the trial receive the treatment that is observed to be superior (i.e. early ECV if $b_1 > 0$ or late ECV is $b_1 \leq 0$), then $N \simeq$ 1 000 000. Based on a total budget for the planned trial of $2 836 000, it is assumed that the fixed cost of setting up the trial is $C_f = \$500\,000$ and the cost per patient is $C_v = \$1600$.

Using these values for b_0, v_0, σ^2, N, C_f and C_v, the sample size that optimizes ENG(n) can be determined. Optimal sample size is 345 per arm and corresponds to a financial cost of $1 604 000 and an opportunity loss of $23 793. The expected net gain is $736 391, yielding a 45% return on investment. The planned sample size of 730 per arm corresponds to a financial cost of $2 836 000 and an opportunity loss of $50 345. The expected net gain is $179 658, yielding a 6.2% return on investment.

ENG(n) is sensitive to the values used for b_0, v_0, σ^2, N, C_f and C_v. Perhaps the most uncertain item is N, since it is difficult to predict the reaction of patients, clinicians and decision makers to the result of an RCT. If N is only 700 000, then $n^* = 264$, with a financial cost of $1 344 800 and an opportunity loss of $18 207. The expected net gain is $69 521, yielding a 5.1% return on investment. By contrast, for $n = 730$, the expected net gain is $-\$741\,488$ with a corresponding return of -26% on investment. Plots of ENG(n) versus n for $N = 1\,000\,000$ and 700 000 are given in Figure 6.7.

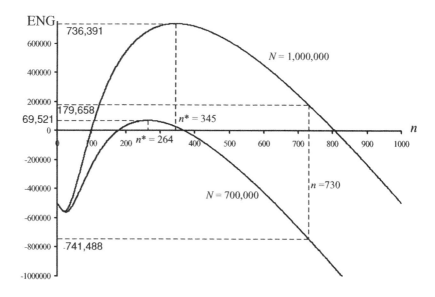

Figure 6.7 Expected net gain as a function of *n* for early ECV trial

6.6 SUMMARY

In this chapter we have tried to give a comprehensive review of sample size determinations for cost-effectiveness trials. Those methods based on inference about the ICER on the cost-effectiveness plane discussed in Section 6.2 tend to be unhelpful, particularly when the solution requires knowledge of the true treatment differences. For inference about the INB we have included a method based on the classical frequentist approach (Section 6.3) and a Bayesian method which allows for the incorporation of prior beliefs (Section 6.4). Finally, we included an INB method based on decision theory using the expected value of information, which provides the sample size that maximizes the difference between the value of the information provided by the trial and the cost of conducting it.

Covariate Adjustment and Sub-group Analysis

7.1 INTRODUCTION

It is often important to adjust for baseline covariates, such as age, sex and disease severity, when estimating Δ_e and Δ_c. A covariate is any variable that influences the patient's measure of effectiveness or cost that is not directly affected by which treatment group the patient is in. Covariates are of a particular concern in observational studies, where one cannot expect the covariate distributions to be the same in both groups, and covariate adjustment might be required to correct potential confounding. A covariate is a confounder if its distribution differs between groups. In randomized clinical trials, groups are expected to have similar distributions with respect to covariates, although bad luck can, and does, produce clinical trials with important differences. Even in the absence of confounding, however, covariate adjustment in clinical trials can provide an increase in the precision of the estimators of Δ_e and Δ_c by removing between-patient variability due to the covariates (see Altman, 1985, Pocock, 1984). Also, through the addition of interaction terms, covariate adjustment can provide analyses of sub-groups defined by the covariates.

In Section 7.2 a system of seemingly unrelated regression equations is used to provide covariate-adjusted estimators of Δ_e and Δ_c for non-censored data (see Greene, 1993; Willan, Briggs and Hoch, 2004). The models can accommodate different sets of covariates for

Statistical Analysis of Cost-effectiveness Data. A. Willan and A. Briggs
© 2006 John Wiley & Sons, Ltd.

effectiveness and costs, but are restricted to measures of effectiveness that are continuous or, at least, ordinal in nature. In these models, treatment indicators are added as covariates, the coefficients of which are the parameters of interest Δ_e and Δ_c. In addition, the use of the models to conduct sub-group analyses is demonstrated. Similar models are described in Section 7.4 for censored data (see Willan, Lin and Manca, 2005). Examples using non-censored data are given in Section 7.3 and for censored data in Section 7.5.

7.2   NON-CENSORED DATA

Dropping the index for treatment arm, let E_i and C_i be the observed effectiveness and cost for patient i, and let $t_i = 1$ if patient i is randomized to T, and 0 if randomized to S. Therefore, $\Delta_e = E(E_i|t_i = 1) - E(E_i|t_i = 0)$ and $\Delta_c = E(C_i|t_i = 1) - E(C_i|t_i = 0)$. If the measure of effectiveness is survival time, then $E_i = X_{ji}$ (see Section 2.3.2) and if the measure of effectiveness is quality-adjusted survival time, then $E_i = q_{ji}$ (see Section 2.3.3)

For patient i, consider the following regression model:

$$E_i = \Delta_e t_i + \omega_0 + \omega_1 z_{1i} + \omega_2 z_{2i} + \ldots \omega_{p_e} z_{p_e i} + \varepsilon_{ei} \qquad \text{and}$$
$$C_i = \Delta_c t_i + \theta_0 + \theta_1 w_{1i} + \theta_2 w_{2i} + \ldots \theta_{p_c} w_{p_c i} + \varepsilon_{ci}$$

where $z_{1i}, z_{2i}, \ldots z_{p_e i}$ are the p_e covariates for effectiveness, $w_{1i}, w_{2i}, \ldots w_{p_c i}$ are the p_c covariates for cost, and $(\varepsilon_{ei}, \varepsilon_{ci})^T$ are independent and identically distributed with mean $(0, 0)^T$ and covariance $\Sigma = \begin{pmatrix} \sigma_e^2 & \sigma_{ec} \\ \sigma_{ec} & \sigma_c^2 \end{pmatrix}$. In matrix form, the models can be written as

$$\mathbf{e} = Z\omega + \varepsilon_e \qquad \text{and}$$
$$\mathbf{c} = W\theta + \varepsilon_c$$

where $E = (E_1, \ldots E_{n_T+n_S})^T$, the ith row of $Z = (t_i, 1, z_{1i}, \ldots z_{p_e i})$, $\omega = (\Delta_e, \omega_0, \omega_1, \ldots \omega_{p_e})^T$, $\varepsilon_e = (\varepsilon_{e1}, \ldots \varepsilon_{e,n_T+n_S})^T$, $\mathbf{c} = (C_1, \ldots C_{n_T+n_S})^T$, the ith row of $W = (t_i, 1, w_{1i}, \ldots w_{p_c i})$, $\theta = (\Delta_c, \theta_0, \theta_1, \ldots \theta_{p_c})^T$, and $\varepsilon_c = (\varepsilon_{c1}, \ldots \varepsilon_{c,n_T+n_S})^T$.

Following Greene (1993) and Willan, Briggs and Hoch (2004), the models for effectiveness and cost can be written as a system of seemingly unrelated regression equations as

$$y = X\beta + \varepsilon$$

where

$$y = \begin{pmatrix} e \\ c \end{pmatrix} \qquad X = \begin{pmatrix} Z & 0 \\ 0 & W \end{pmatrix} \qquad \beta = \begin{pmatrix} \omega \\ \theta \end{pmatrix} \qquad \varepsilon = \begin{pmatrix} \varepsilon_e \\ \varepsilon_c \end{pmatrix}$$

and 0 represents matrices of appropriate dimensions in which all elements are zero. The vector of parameters β can be estimated by the ordinary least-squares (OLS) solution given by

$$\hat{\beta}_{ols} = \left(X^T X \right)^{-1} X^T y$$

with variance–covariance matrix given by

$$V(\hat{\beta}_{ols}) = \left[X^T \left(\Sigma^{-1} \otimes I \right) X \right]^{-1},$$

where I is the identity matrix of dimension $n_T + n_S$. The vector of residuals is given by

$$\hat{\varepsilon} = \left[I - X(X^T X)^{-1} X^T \right] y$$

and the elements of Σ can be consistently estimated by

$$\hat{\sigma}_e^2 = \sum_{i=1}^{n_T+n_S} \hat{\varepsilon}_i \hat{\varepsilon}_i / [n_T + n_S - (p_e + 2)]$$

$$\hat{\sigma}_c^2 = \sum_{i=n_T+n_S+1}^{2(n_T+n_S)} \hat{\varepsilon}_i \hat{\varepsilon}_i / [n_T + n_S - (p_c + 2)]$$

$$\hat{\sigma}_{ce} = \sum_{i=1}^{n_T+n_S} \hat{\varepsilon}_i \hat{\varepsilon}_{n_T+n_S+i} \Big/ \left[n_T + n_S - \max_{j=T,S}(p_j + 2) \right]$$

and $\hat{\varepsilon}_i$ is the ith component of $\hat{\varepsilon}$.

If the covariates for effectiveness and cost are the same (i.e. $U = W$), then the OLS estimator is the best linear unbiased estimator (BLUE). If one set of covariates is a subset of the other, then the OLS estimator for the smaller equation is BLUE. In all other situations, depending on the data, efficiency gains are possible by using the generalized least squares (GLS) solution, given by

$$\hat{\beta}_{gls} = \left[X^T (\Sigma^{-1} \otimes I) X \right]^{-1} X^T (\Sigma^{-1} \otimes I) \mathbf{y} \qquad (7.1)$$

with covariance given by

$$V(\hat{\beta}_{gls}) = \left[X^T (\Sigma^{-1} \otimes I) X \right]^{-1} \qquad (7.2)$$

where $\otimes$ is the Kronecker (direct) product. For Equation (7.1), Σ can be estimated from the OLS solution as shown above. For Equation (7.2), the elements of Σ are estimated using the residuals for the GLS solution. Whether or not the GLS solution provides efficiency gains depends on the data. In general, the efficiency gains increase as the absolute value of $\rho = \sigma_{ec}/\sigma_e\sigma_c$, the correlation between effectiveness and cost, increases. Also, the efficiency gains increase as the independence between the set of regression variables for cost and the set of regression variables for effectiveness increases. Efficiency gains will be reduced if the treatment variable is correlated with the covariates for either cost or effectiveness (confounding) and if the covariates for cost and the covariates for effectiveness are correlated.

If we define $\hat{\beta} = \hat{\beta}_{ols}$ or $\hat{\beta}_{gls}$, depending on which solution is being used, then

$$\hat{\beta} = \left(\hat{\Delta}_e, \hat{\omega}_0, \ldots \hat{\omega}_{p_e}, \hat{\Delta}_c, \hat{\theta}_0, \ldots \hat{\theta}_{p_e} \right)^T$$

Therefore, the first element of $\hat{\beta}$ is the covariate-adjusted estimator of Δ_e and the $(p_e + 3)$th element is the covariate-adjusted estimator of Δ_c. Let $v_{mk} =$ the mth, kth element of $\left[X^T (\hat{\Sigma}^{-1} \otimes I) X \right]^{-1}$. Then

$$\hat{V} \left(\hat{\Delta}_e \right) = v_{1,1}$$
$$\hat{V} \left(\hat{\Delta}_c \right) = v_{p_e+3, \, p_e+3} \qquad \text{and}$$
$$\hat{C} \left(\hat{\Delta}_e, \hat{\Delta}_c \right) = v_{1, p_e+3}$$

Thus, we have the covariate-adjusted estimators for the five parameters required to perform a cost-effectiveness analysis as described in Chapter 4, using either an ICER, INB or CEAC approach.

The regression model allows for sub-group analysis. Suppose investigators were interested in determining whether a certain sub-group, say males for sake of argument, have the same incremental net benefit as females. To accomplish this let z_{1i} be 1 if the ith patient is male and 0 if female, and let z_{2i} equal $t_i z_{1i}$. The covariate u_{2i} is the interaction between sex and treatment arm. Let $w_{1i} = z_{1i}$ and $w_{2i} = z_{2i}$. There could be any number of additional covariates for either effectiveness and cost other than these. The parameter of interest is $\lambda \omega_2 - \theta_2$. If $\lambda \hat{\omega}_2 - \hat{\theta}_2$ is positive, then males (since they are coded 1) are observed to have greater INB(λ). The opposite is true if $\lambda \hat{\varphi}_2 - \hat{\theta}_2$ is negative. The hypothesis that $\text{INB}_{\text{males}}(\lambda) = \text{INB}_{\text{females}}(\lambda)$ versus $\text{INB}_{\text{males}}(\lambda) \neq \text{INB}_{\text{females}}(\lambda)$ can be rejected at the two-sided, level α, if $\left|\lambda \hat{\omega}_2 - \hat{\theta}_2\right| \big/ \sqrt{\lambda^2 v_{44} + v_{p_e+4, p_e+4} - 2\lambda v_{4, p_e+4}}$ exceeds $z_{1-\alpha/2}$, where p_e now is the number of covariates plus the number of any interaction terms, i.e. the number of 'z-terms' in the regression model. The expression $\lambda^2 v_{44} + v_{p_e+4, p_e+4} - 2\lambda v_{4, p_e+4}$ is the variance of $\lambda \hat{\omega}_2 - \hat{\theta}_2$. When testing for interactions both terms should be included. Observing that $\hat{\omega}_2, \hat{\theta}_2$ or both, are not statistically significant does not mean that they are known to be zero with 100% certainty, and the additional test of the hypothesis $\lambda \omega_2 - \theta_2 = 0$ is required to determine if there is a significant interaction between the variable in question and treatment arm with respect to INB.

7.2.1 Example, non-censored data

A randomized controlled trial was conducted to compare an assertive community treatment (ACT) program with usual community services in patients with severe and persistent mental illnesses (SPMI). A brief overview is given here; for further details and a complete report of the results the reader is referred to Lehman *et al.* (1997; 1999). The measure of effectiveness is the number of days in stable housing. A day of stable housing was defined as living in a uninstitutionalised setting not intended to serve the homeless (e.g. independent housing, living with family, etc.). Costs are given in US dollars. The duration of interest was 1 year. A total of 148 persons who were homeless with SPMI were randomized to either the experimental ACT program (T) or to usual community services (S). Subjects were recruited during a 19-month period in 1991 and 1992 from Baltimore inner-city

Table 7.1 Subject characteristics by treatment arm

Characteristic	$ACT^a (n = 73)$	Usual services $(n = 72)$
Age (years): mean (SD)	39.0 (9.43)	36.0 (8.30)
GAF^b Score: mean (SD)	37.9 (9.08)	35.3 (9.06)
African American[*] (%)	62	83

[a] Assertive community treatment.
[b] Global assessment of functioning.
[*] $p < 0.01$.

psychiatric hospitals, primary health care agencies, shelters, missions and soup kitchens. Baseline data collection included age, race and an assessment of overall mental health functioning using the global assessment of functioning (GAF) scale.

Complete data were collected on 73 participants assigned to ACT (T) and 72 assigned to usual services (S). The comparison between treatment arms with respect to the baseline variables is given in Table 7.1. The two arms appear comparable with respect to age and GAF scores. However, a greater percentage of subjects randomized to usual services were African Americans and, therefore, race is a potential confounder. The sample mean effectiveness and cost, broken down by treatment arm and race, are given in Table 7.2 along with the between-treatment differences. ACT subjects had lower average costs and higher average number days of stable housing, suggesting it is the dominant treatment. The ACT program is observed to have a greater effect among White subjects in whom there is a larger decrease in average cost ($62 700 vs $5070) and a larger increase in average effect (98.1 vs 35.6 days).

The observed interaction together with the fact that there was a disproportionate number of African Americans in the usual care group was the motivation for fitting a regression model, including an indicator variable for race and its interaction with randomization group (Model 1). The indicator variable for race is I{patient is African American}, and the indicator for randomization group is I{patient randomized to ACT}. The results are shown in Table 7.3. Because the coefficient for race and its interaction were significant for cost there

Table 7.2 Simple statistics

	Mean	Standard error
ACT subjects		
Cost	51 880	7156
Effectiveness	211.9	12.22
Comparison subjects		
Cost	67 400	9020
Effectiveness	159.2	12.41
All subjects		
Cost difference	−15520	11 500
Effectiveness difference	52.67	17.42
Stratified analysis		
African American subjects		
Cost difference	−5072	13 110
Effectiveness difference	35.61	21.34
White subjects		
Cost difference	−62750	25 070
Effectiveness difference	98.1	32.68

Table 7.3 Parameter estimates for regression Model 1

Coefficient [standard error] (2-sided *p*-value)	Effectiveness	Cost
Randomization group	98.10 [36.14] (0.006643)	−62, 750 [23 530] (0.007665)
Intercept	132.7 [30.24] (0.00001154)	112 200 [19 690] (1.193×10^{-8})
Race	31.92 [33.12] (0.3353)	−53810 [21,570] (0.01260)
Race*Randomization group	−62.48 [41.63] (0.1334)	57 680 [27,100] (0.03334)

Table 7.4 Cost-effectiveness parameter estimates for regression Model 1

	African Americans	Whites
$\hat{\Delta}_e$	35.61	98.1
$\hat{\Delta}_c$	−5072	−62750
$\hat{V}(\hat{\Delta}_e)$	426.7	1306
$\hat{V}(\hat{\Delta}_c)$	1.809×10^8	5.538×10^8
$\hat{C}(\hat{\Delta}_e, \hat{\Delta}_c)$	-1.119×10^5	-3.425×10^5

is some evidence for concluding that the treatment effect for cost depends on race; the implication being that a separate cost-effectiveness analysis is required for each race. The cost-effectiveness parameter estimates for each race are shown in Table 7.4. The main effect parameter estimates from regression Model 1 are those applicable to Whites because they are coded 0. The simplest way to get the parameter estimates for African Americans is to reverse the coding (i.e. Whites = 1 and African Americans = 0) and rerun the regression procedure. From the parameter estimates in Table 7.4, the estimate for the ICER for African Americans is −142.4 with 90% confidence limits −1213 and 3363; the corresponding values for Whites are −639.7, −1439 and −278.4. Since the confidence interval falls entirely within the dominant quadrant for Whites, it can be concluded with a high degree of confidence that the treatment is cost-effective for them regardless of the willingness-to-pay for a homeless day. However, for African Americans, although treatment is observed to dominate, the upper confidence limit of $3363 per homeless day implies that conclusive evidence for cost-effectiveness is limited to willingness-to-pay values in excess of this value. These conclusions can be seen also in the INB plots in Figure 7.1. For clarity, and because they are of secondary interest, the upper limits of INB are not shown. Since treatment is observed to dominate in both racial groups, INB(λ) is positive regardless of λ. For Whites the lower limit is always positive, implying that the hypothesis INB(λ) $\leq$ 0 can be rejected for any value of λ. For African Americans the lower limit crosses the horizontal axis at $3363, and the hypothesis INB(λ) $\leq$ 0 cannot be rejected for $\lambda \leq 3363$ at the

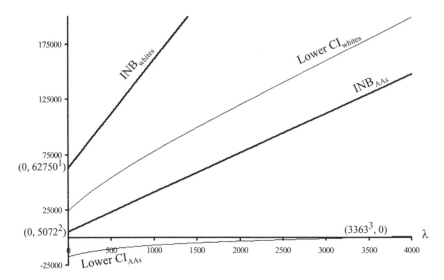

Figure 7.1 INB and lower 90% confidence limit versus lambda, Model 1. [1] $-\hat{\Delta}_c$ for whites; [2] $-\hat{\Delta}_c$ for African Americans; [3] ICER upper confidence limit for African Americans

5% level. (The use of 90% confidence intervals provides for 5%, one-sided test of hypotheses.)

Since the terms for race and race by treatment are not significant for effectiveness, one may be tempted to fit a model without those two terms (Model 2). The results are shown in Table 7.5 for OLS and GLS. The standard errors for the parameters for costs are slightly smaller for GLS, implying relatively modest efficiency gains. The efficiency gains are modest, even though the absolute value for ρ of 0.40 is large, because the set of regression variables for cost and the set of regression variables for effectiveness are correlated. First, for obvious reasons, the treatment variable is in both sets. Second, race, which is part of the regression equation for cost, but not effectiveness, is confounded with the treatment variable, see Table 7.1. Efficiency gains will often be modest in this context, and an important motivation for using the method of seemingly unrelated regression equations is to provide an estimate of $C(\hat{\Delta}_e, \hat{\Delta}_c)$, which is required for statistical inference in cost-effectiveness analyses.

Table 7.5 Parameter estimates for regression Model 2

Coefficient [standard error] (2-sided p-value)	Effectiveness	Cost	
	OLS/GLS	OLS	GLS
Randomization group	52.66 [17.42] (0.002508)	−62 750 [22 020] (0.004381)	−50 870 [22 000] (0.02079)
Intercept	159.2 [12.36] (0.00001154)	112 200 [18 310] (8.866×10^{-10})	105 300 [18 290] (8.658×10^{-9})
Race		−53 810 [19 750] (0.006432)	−45 460 [19 710] (0.02110)
Race*Randomization group		57 680 [24 820] (0.02012)	41 340 [24 780] (0.09520)

The cost-effectiveness parameter estimates using GLS for this model are shown in Table 7.6. From these parameter estimates the estimate for the ICER for African Americans is −180.9 with 90% confidence limits −608 and 317; the corresponding values for Whites are −966, −2227 and −294. The INB plots for Model 2 can be found in

Table 7.6 GLS cost-effectiveness parameter estimates for regression Model 2

	African Americans	Whites
$\hat{\Delta}_e$	52.66	52.66
$\hat{\Delta}_c$	−9528	−50870
$\hat{V}(\hat{\Delta}_e)$	303.5	303.5
$\hat{V}(\hat{\Delta}_c)$	1.726×10^8	4.841×10^8
$\hat{C}(\hat{\Delta}_e, \hat{\Delta}_c)$	-8.071×10^4	-8.071×10^4

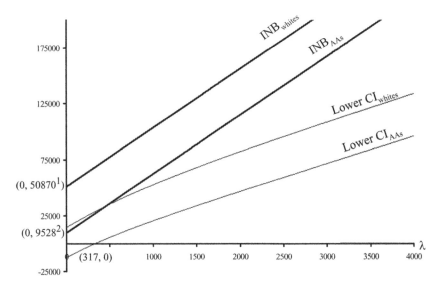

Figure 7.2 INB and lower 90% confidence limit versus lambda, Model 2. [1]$-\hat{\Delta}_c$ for whites; [2]$-\hat{\Delta}_c$ for African Americans; [3]ICER upper confidence limit for African Americans

Figure 7.2. As with Model 1, the lower confidence limit for Whites is positive for all values of λ, and it can be concluded that the treatment is cost-effective for them regardless of the willingness-to-pay for a homeless day. For African Americans, the upper confidence limit for the ICER of \$317 per homeless day implies that conclusive evidence for cost-effectiveness is limited to willingness-to-pay values in excess of this value. However, this threshold value for the willingness-to-pay is less than a tenth of the value for Model 1. Consequently, the health policy implications of the results appear very sensitive to the inclusion of two non-significant parameters in the regression model. The contrast in the ICER upper limits is better illustrated in Figure 7.3 in which the CEAC for African Americans is plotted for both models. In fact, it is the arbitrary use of 95% confidence that is responsible in large part for the apparent difference in the two models. For Model 2, the probability that INB is positive is 95% for a WTP of \$317. For Model 1, the comparable WTP is \$3364, more than ten times the value for Model 2. Nonetheless, for a WTP of \$317, the probability that INB is positive for Model 1 is 83%, which is only 12 percentage points less

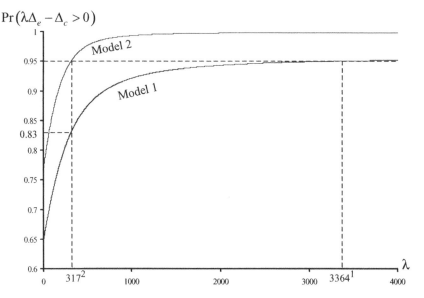

Figure 7.3 CEACs for African Americans. [1]ICER upper 90% confidence limit for African Americans, model 1; [2]ICER upper 90% confidence limit for African Americans, model 2

than for Model 2. By looking at the entire CEAC, rather than just one arbitrary point, one would see evidence for cost-effectiveness using either model.

Nonetheless, it is Model 1 that must be used for testing interactions on the INB scale. As discussed in Section 7.2, when testing interactions to identify sub-groups for which the INB differs significantly, the appropriate covariates and their interactions for both effectiveness and cost must be included. That is, the relevant test of hypothesis is $\lambda\omega_2 - \theta_2 = 0$. The fact that $\hat{\omega}_2$ is not significantly different from zero does not mean that we know it is zero with 100% confidence. The p-values for various values of λ using the parameter estimates from Model 1, are given in Table 7.7.

Finally, a model was fitted considering all three covariates (Model 3). The results are presented in Table 7.8. The negative coefficients for the interaction terms for effectiveness indicates that the benefit of ACT was observed to be greater for Whites, younger subjects and

Table 7.7 p-values as a function of λ for testing H: $\omega_2 \lambda - \theta_2 = 0$ in regression Model 1

	p-value*
0 (costs)	0.03334
50	0.02994
100	0.02767
200	0.02541
500	0.02727
1000	0.03864
2000	0.06073
5000	0.09320
∞ (effectiveness)	0.1334

subjects with lower GAF scores. On the cost side, the coefficients for the interaction terms are all positive, indicating that the reduction in costs from ACT were observed to be greater for Whites, younger subjects and subjects with lower GAF scores. Taken together, the signs of the interaction terms indicate that for any given λ the INB(λ) is observed to be greater for Whites, younger subjects and subjects with lower GAF scores. The p-values for the test of hypothesis for the INB(λ) interactions are given in Table 7.9. For many plausible values of λ there is a significant interaction between randomization group and each of the three covariates. Note that for GAF, neither $\hat{\omega}_6$ nor $\hat{\theta}_6$ are significant; however, for many values of λ, $\lambda\hat{\omega}_6 - \hat{\theta}_6$ is significant. The presence of the interactions makes summarizing the results difficult. Even if age and GAF scores were dichotomized, eight INB(λ) plots are required to illustrate the results.

7.3   CENSORED DATA

Adjusting for covariates is somewhat more complex for censored data. The use of inverse-probability weighting for covariate adjustment is illustrated for cost in Section 7.4.1, for quality-adjusted survival time in Section 7.4.2, and for survival time in Section 7.4.3. The respective

Table 7.8 Parameter estimates for regression Model 3

Coefficient [standard error] (2-sided p-value)	Effectiveness	Cost
	OLS/GLS	OLS/GLS
Randomization group	298.1 [105.4] (0.004669)	−255 000 [68,850] (0.0002125)
Intercept	−67.10 [70.37] (0.3404)	204 700 [45 990] (8.507×10^{6})
Race	32.06 [32.73] (0.3273)	−54 330 [21 390] (0.01108)
Age	2.608 [1.531] (0.08842)	−1319 [1000] (0.1872)
GAF	2.994 [1.400] (0.03254)	−1262 [915.2] (0.1680)
Race*Randomization group	−61.53 [41.16] (0.1349)	60 210 [26 900] (0.02519)
Age*Randomization group	−2.368 [2.008] (0.2382)	2861 [1312] (0.02920)
GAF*Randomization group	−32.64 [1.941] (0.09270)	2276 [1269] (0.07282)

estimator for covariance between $\hat{\Delta}_e$ and $\hat{\Delta}_c$ is also given in Sections 7.4.2 and 7.4.3. The methods are similar to those proposed by Lin (2000) and Willan, Lin and Manca (2005). Examples are given in Section 7.5.

Table 7.9 *p*-values for testing INB interactions in regression Model 3 as a function of λ

λ	Race H: $\omega_4\lambda - \theta_4 = 0$	Age H: $\omega_5\lambda - \theta_5 = 0$	GAF H: $\omega_6\lambda - \theta_6 = 0$
0 (costs)	0.02519	0.02920	0.07282
50	0.02255	0.02769	0.06234
100	0.02083	0.02698	0.05468
200	0.01924	0.02731	0.04501
500	0.02158	0.03622	0.03639
1000	0.03297	0.06041	0.03965
2000	0.05581	0.1041	0.05143
5000	0.09070	0.1658	0.06981
∞(effectiveness)	0.1349	0.2382	0.09270

7.3.1 Cost

Again dropping the index for treatment, and using similar notation as in Chapter 3, for patient i let D_i and U_i be the times from randomization to death and censoring, respectively. Let $G(t) = \Pr(U_i \geq t)$, $X_i = \min(D_i, U_i)$, $\delta_i = I\{D_i \leq U_i\}$ and $\bar{\delta}_i = 1 - \delta_i$. The function $G(t)$ is the probability of not being censored in the interval $[0, t)$. The time period of interest $(0, \tau]$ is again divided in to K intervals $(a_k, a_{k+1}]$, $k = 1, 2, \ldots K$, where $0 = a_1 < a_2 < \ldots a_{K+1} = \tau$. Typically the intervals $(a_k, a_{k+1}]$ are the shortest intervals for which disaggregated cost data are available, and usually correspond to the intervals between data collection visits. If cost histories are not available, and cost data are aggregated for each patient over the entire duration of interest, then $K = 1$, $a_1 = 0$ and $a_{K+1} = \tau$. Let C_{ki} be the observed cost for patient i during interval k. Naturally, $C_{ki} = 0$ for any interval that begins after patient i dies, i.e. if $a_k \geq D_i$. Let $X_{ki}^* = \min(X_i, a_{k+1})$ and $C_i = \sum_{k=1}^{K} C_{ki}$. Consider the linear regression model $E(C_i) = \theta^T W_i$, where W_i is the column vector whose components constitute the ith row of the matrix W, as defined in Section 7.2, where the first component is patient i's indicator variable for treatment arm, the second component is 1 and the remaining components are the covariate values for patient i. The first component of θ is the parameter of primary interest, Δ_c, the difference between treatments in mean cost.

From Lin (2000) θ can be estimated by $\hat{\theta} = \sum_{k=1}^{K} \hat{\theta}_k$, where

$$\hat{\theta}_k = \left(\sum_{i=1}^{n} \frac{\delta_{ki}^*}{\hat{G}(X_{ki}^*)} W_i W_i^T \right)^{-1} \sum_{i=1}^{n} \frac{\delta_{ki}^* C_{ki}}{\hat{G}(X_{ki}^*)} W_i$$

where $\hat{G}(t)$ is the product-limit estimator of $G(t)$, and $\delta_{ki}^* = \delta_i + \bar{\delta}_i I\{X_i \geq a_{k+1}\}$, i.e. δ_{ki}^* is 1 if the patient dies or if the patient is censored after the end of interval k. The first component of $\hat{\theta}$ is $\hat{\Delta}_c$, the estimator for the difference between treatments in mean cost, adjusted for the covariates. The covariance matrix for $\hat{\theta}$ is estimated by

$$\hat{V}\left(\hat{\theta}\right) = n^{-1} \hat{A}_C^{-1} \hat{B}_C \hat{A}_C^{-1}$$

where

$$\hat{A}_C = n^{-1} \sum_{i=1}^{n} W_i W_i^T, \qquad \hat{B}_C = n^{-1} \sum_{i=1}^{n} \sum_{k=1}^{K} \sum_{m=1}^{K} \hat{\xi}_{ki}^{(c)} \hat{\xi}_{mi}^{(c)T}$$

$$\hat{\xi}_{ki}^{(c)} = \left(\frac{\delta_{ki}^*(C_{ki} - \hat{\theta}_k^T W_i)}{\hat{G}(X_{ki}^*)} W_i + \bar{\delta}_i F_{Cki} - \sum_{j=1}^{n} \frac{\bar{\delta}_j I\{X_j \leq X_i\}}{\mathbf{R}_j} F_{Ckj} \right)$$

$$R_i = \sum_{j=1}^{n} I\{X_j \geq X_i\} \qquad \text{and}$$

$$F_{Cki} = \frac{1}{R_i} \sum_{j=1}^{n} \frac{I\{X_{kj}^* > X_i\} \delta_{kj}^* \left(C_{kj} - \hat{\theta}_k^T W_j \right)}{\hat{G}(X_{kj}^*)} W_j$$

and n is the total sample size $n_T + n_S$. The element in the first row and first column of $\hat{V}\left(\hat{\theta}\right)$ is the estimator of the variance of $\hat{\Delta}_c$.

7.3.2 Quality-adjusted survival time

Let q_{ki} be the observed quality of life experienced by the patient i during interval k, determined as described in Section 3.3.3. Let $q_i = \sum_{k=1}^{K} q_{ki}$

and consider the linear regression model $E(q_i) = \omega^T Z_i$, where Z_i is the column vector whose components constitute the ith row of the matrix Z, as defined in Section 7.2, where the first component is patient i's indicator variable for treatment arm, the second component is 1 and the remaining components are the covariate values for patient i. The first component of ω is the parameter of primary interest, Δ_e, the difference between treatments in mean quality-adjusted survival. The estimation procedure is the same as that given in Section 7.4.1 for cost, only with q_{ki} substituted for C_{ki}.

Thus ω is estimated by $\hat{\omega} = \sum_{k=1}^{K} \hat{\omega}_k$

where

$$\hat{\omega}_k = \left(\sum_{i=1}^{n} \frac{\overset{*}{\delta}_{ki}}{\hat{G}(\overset{*}{X}_{ki})} Z_i Z_i^T \right)^{-1} \sum_{i=1}^{n} \frac{\overset{*}{\delta}_{ki} q_{ki}}{\hat{G}(\overset{*}{X}_{ki})} Z_i$$

The first component of $\hat{\omega}$ is $\hat{\Delta}_e$, the estimator for the difference between treatments in mean quality-adjusted survival time, adjusted for the covariates. The covariance matrix for $\hat{\omega}$ is estimated by

$$\hat{V}(\hat{\omega}) = n^{-1} \hat{A}_E^{-1} \hat{B}_E \hat{A}_E^{-1}$$

where

$$\hat{A}_E = n^{-1} \sum_{i=1}^{n} Z_i Z_i^T, \qquad \hat{B}_E = n^{-1} \sum_{i=1}^{n} \sum_{k=1}^{k} \sum_{m=1}^{k} \hat{\xi}_{ki}^{(e)} \hat{\xi}_{mi}^{(e)T}$$

$$\hat{\xi}_{ki}^{(e)} = \left(\frac{\overset{*}{\delta}_{ki}(q_{ki} - \hat{\omega}_k^T Z_i)}{\hat{G}(\overset{*}{X}_{ki})} Z_i + \bar{\delta}_i F_{Eki} - \sum_{j=1}^{n} \frac{\bar{\delta}_j I\{X_j \le X_i\}}{R_j} F_{Ekj} \right)$$

$$R_i = \sum_{j=1}^{n} I\{X_j \ge X_i\} \qquad \text{and} \quad F_{Cki}$$

$$= \frac{1}{R_i} \sum_{j=1}^{n} \frac{I\{\overset{*}{X}_{kj} > X_i\} \overset{*}{\delta}_{kj} (q_{kj} - \hat{\omega}_k^T Z_j)}{\hat{G}(\overset{*}{X}_{kj})} Z_j$$

The element in the first row and first column of $\hat{V}(\hat{\omega})$ is the estimator of the variance of $\hat{\Delta}_e$. The matrix of covariances between

$\hat{\theta}$ and $\hat{\omega}$ is estimated by $\hat{C}(\hat{\theta}, \hat{\omega}) = n^{-1} \hat{A}_C^{-1} \hat{B}_{CE} \hat{A}_E^{-1}$, where. $\hat{B}_{CE} = n^{-1} \sum_{i=1}^n \sum_{k=1}^k \sum_{m=1}^k \hat{\xi}_{ki}^{(c)} \hat{\xi}_{mi}^{(e)T}$. The element in the first row and first column of $\hat{C}(\hat{\theta}, \hat{\omega})$ is the covariance between $\hat{\Delta}_c$ and $\hat{\Delta}_e$.

7.3.3 Survival time

Let $D_i^* = \min(D_i, \tau)$ and consider the linear regression model $E(D_i^*) = \omega^T Z_i$, where w and Z_i are defined in Section 7.4.2. As shown in Willan, Lin and Manca (2005), can be estimated by

$$\hat{\omega} = \left(\sum_{i=1}^n \frac{\delta_i^*}{\hat{G}(X_i^*)} Z_i Z_i^T \right)^{-1} \sum_{i=1}^n \frac{\delta_i^* X_i}{\hat{G}(X_i^*)} Z_i$$

where $X_i^* = \min(X_i, \tau)$ and $\delta_i^* = \delta_i + \bar{\delta}_i I(U_i \geq \tau)$. The first component of $\hat{\omega}$ is $\hat{\Delta}_e$, the estimator for the difference between treatments in mean survival time, adjusted for the covariates. The covariance matrix for $\hat{\omega}$ is estimated by

$$\hat{V}(\hat{\omega}) = n^{-1} \hat{A}_E^{-1} \hat{B}_E \hat{A}_E^{-1}$$

where

$$\hat{A}_E = n^{-1} \sum_{i=1}^n Z_i Z_i^T, \qquad \hat{B}_E = n^{-1} \sum_{i=1}^n \hat{\xi}_i^{(e)} \hat{\xi}_i^{(e)T}$$

$$\hat{\xi}_i^{(e)} = \left(\frac{\delta_i^* (X_i - \hat{\omega}^T Z_i)}{\hat{G}(X_i^*)} Z_i + \bar{\delta}_i F_{Ei} - \sum_{j=1}^n \frac{\bar{\delta}_j I\{X_j \leq X_i\}}{R_j} F_{Ej} \right) \quad \text{and}$$

$$F_{Ei} = \frac{1}{R_i} \sum_{j=1}^n \frac{I\{X_j^* > X_i\} \delta_j^* (X_j - \hat{\omega}^T Z_j)}{\hat{G}(X_j^*)} Z_j.$$

The element in the first row and first column of $\hat{V}(\hat{\omega})$ is the estimator of the variance of $\hat{\Delta}_e$. The matrix of covariances between $\hat{\theta}$ and $\hat{\omega}$ is estimated by $\hat{C}(\hat{\theta}, \hat{\omega}) = n^{-1} \hat{A}_C^{-1} \hat{B}_{CE} \hat{A}_E^{-1}$, where $\hat{B}_{CE} = n^{-1} \sum_{i=1}^n \left(\sum_{k=1}^K \hat{\xi}_{ki}^{(c)} \right) \hat{\xi}_i^{(e)T}$. The element in the first row and first column of $\hat{C}(\hat{\theta}, \hat{\omega})$ is the covariance between $\hat{\Delta}_c$ and $\hat{\Delta}_e$.

7.3.4   The Canadian implantable defibrillator study (CIDS)

For the CIDS trial introduced in Section 5.4, consider the covariates left ventricular ejection fraction (F) and sex (X). $F = I\{$patient's ejection fraction $\leq 35\%\}$ and $X = I\{$patient is male$\}$. The treatment variable is coded 1 for defibrillator and 0 for amiodarone. The number and proportion of males and those whose ejection fraction was equal to or less than 35% is shown in Table 7.10. For any particular regression model, we used only those observations for which there were no missing data for the covariates included. Using survival time as the measure of effectiveness, the parameter estimates, standard errors and p-values are displayed in Table 7.11, for the model with no covariates, the model with each covariate alone and the model with both covariates. Costs are given in Canadian dollars.

Ejection fraction was statistically significant for both cost and effectiveness, whether or not sex was included in the model. Sex was not significant for cost or effectiveness, regardless of whether ejection fraction was included in the model. The signs for the ejection fraction coefficients were positive for cost and negative for effectiveness, indicating that patients with ejection fraction equal to or less than 35% cost on average \$9012 more and lived an average of 0.8974 years less. Because of this and the fact that there were slightly more patients in the amiodarone arm whose ejection fraction was equal to or less than

Table 7.10   Covariates by treatment, CIDS example

		Treatment	
		Defibrillator ($n_T = 212$)	Amiodarone ($n_S = 218$)
Ejection fraction* $\leq 35\%$	n	132	147
	(%)	64.7	68.7
Males	n	184	174
	(%)	86.8	79.8

* 12 missing values.

Table 7.11   Parameter estimates, standard errors and p-values, CIDS example

Parameter		No covariates	Ejection fraction alone	Sex alone	Ejection fraction and sex
$\hat{\Delta}_c$	Estimate	48 668	49 666	48 488	49 586
	SE	3 956	3 997	4 083	4 120
	p-value	< 0.0001	< 0.0001	< 0.0001	< 0.0001
$\hat{\theta}_F$	Estimate		9012		8 857
	SE	n.i.m.	3871	n.i.m.	3 901
	p-value		0.0199		0.0232
$\hat{\theta}_X$	Estimate			2 876	1 579
	SE	n.i.m.	n.i.m.	5 530	5 579
	p-value			0.6029	0.7771
$\hat{\Delta}_e$	Estimate	0.4536	0.2958	0.4493	0.2912
	SE	0.4131	0.4030	0.4197	0.4012
	p-value	0.2722	0.4629	0.2844	0.4683
$\hat{\omega}_F$	Estimate		-0.8974		-0.9079
	SE	n.i.m.	0.3891	n.i.m.	0.3885
	p-value		0.0211		0.0195
$\hat{\omega}_X$	Estimate			0.3834	0.3999
	SE	n.i.m.	n.i.m.	0.5607	0.5075
	p-value			0.494.	0.4307
$\hat{C}\left(\hat{\Delta}_e, \hat{\Delta}_c\right)$		359.3	365.1	366.4	370.7

n.i.m. not in model.

35%, when ejection fraction is included in the model, the treatment effect increases for cost and decreases for effectiveness. Therefore, the unadjusted ICER is $107 293/life-year which is less than the adjusted ICER equal to $167 904/life-year. Ignoring the possibility of an interaction between treatment and ejection fraction for the moment, one might choose to use the parameter estimates from the model with ejection fraction alone in the cost-effectiveness analysis. The adjusted

90% Fieller confidence limits for the ICER are 52 728 and −130 609. Because the lower limit lies in the NE quadrant and the upper limit lies in the NW quadrant of the cost-effectiveness plane, the confidence interval includes undefined values (the positive vertical axis), indicating that, because $\hat{\Delta}_e$ is 'statistically' close to zero, there is no upper limit to the ICER, and the proper representation of the confidence interval is [52, 728, ∞).

The plot of the adjusted estimate of INB and its 90% confidence limits by λ can be found in Figure 7.4. Minus $\hat{\Delta}_c$ and the corresponding 90% confidence limits are given where the plots meet the vertical axis. The horizontal intercepts provide the estimate of the ICER and its Fieller confidence limits. The 90% lower limit of the INB never crosses the horizontal axis, indicating that no value of the willingness-to-pay would lead us to reject the hypothesis INB(λ) ≤ 0 in favour of INB(λ) at the 5% level. It is for the same reason that the 90% confidence interval for the ICER includes undefined values.

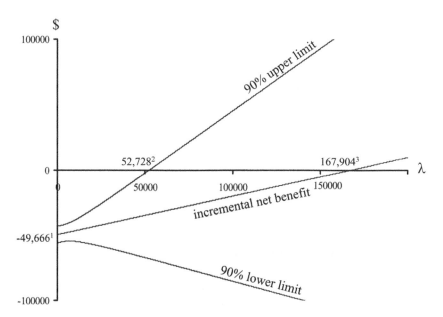

Figure 7.4 INB and confidence limit versus lambda, CIDS example including ejection fraction alone. [1] $\hat{\Delta}_c$; [2] 90% lower limit of ICER; [3] ICER

Table 7.12   INB interaction between treatment and ejection fraction, CIDS example

λ	$\hat{\omega}_{T \times F}\lambda - \hat{\theta}_{T \times F}$	S.E.	p-value
50 000	16 028	37 254	0.667
125 000	37 411	94 578	0.692
250 000	73 050	190 730	0.702
500 000	144 328	383 257	0.706
1000 000	286 885	768 421	0.709

Although not shown in a table, a model was fit which included treatment, ejection fraction and their interaction. The estimated interaction parameter associated with cost was -1772 ($p = 0.818$) and with effectiveness was 0.2851 ($p = 0.711$). The negative estimate for cost implies that the increase in cost due to the defibrillator was observed to be less for patients with a low ejection fraction. The positive estimate for effectiveness implies that the increase in effectiveness due to the defibrillator was observed to be greater for patients with a low ejection fraction. Neither interaction term is significant; however this is not sufficient to conclude that the interaction on the INB scale is non-significant for all values of λ of interest, although in this particular case since the p-values are quite high, it is unlikely that there exist any value of λ for which the interaction is significant on the INB scale. Thus we tested the hypothesis $H : \omega_{T \times F}\lambda - \theta_{T \times F} = 0$, where $T \times F$ indicates the treatment by ejection fraction interaction. The estimate $\hat{\omega}_{T \times F}\lambda - \hat{\omega}_{T \times F}$ its standard error and associated p-value are given in Table 7.12 for several values of λ. The estimate will always be positive for positive values of λ, but for no values in this range was the interaction significant.

7.3.5   The EVALUATE trial

For the EVALUATE trial introduced in Section 5.5, the following variables were considered as potential covariates in the regression models: EQ-5D at randomization (Q), age (A), body mass index (B), smoking status (M), and whether or not the patient had previous pelvic

Table 7.13   Potential continuous covariates by treatment and type of hysterectomy, EVALUATE

		Vaginal			Abdominal		
		Mean	Std. dev.	Median	Mean	Std. dev.	Median
Baseline EQ-5D	Laparoscopic	0.746	0.256	0.760	0.716	0.266	0.760
	Standard	0.758	0.232	0.796	0.691	0.266	0.725
Age	Laparoscopic	41.0	6.83	41.1	41.8	7.00	41.9
	Standard	41.1	6.45	41.0	41.5	7.63	40.9
BMI	Laparoscopic	26.5	5.20	25.5	26.6	5.08	26.0
	Standard	26.6	4.87	26.2	26.1	5.49	24.9

surgery (P). $M = I\{$if the patient smoked$\}$, and $P = I\{$if the patient had previous pelvic surgery$\}$. Descriptive statistics of the covariates by treatment arm are given in Tables 7.13 and 7.14.

The first regression models included the treatment variable alone. The final regression models included a covariate if its corresponding significance level was less than or equal to 0.05 in the model which included the treatment variable and all other significant covariates. The covariate parameter estimates and their corresponding significance levels are given in Table 7.15. A one-year increase in age is observed to increase cost by £24 for a vaginal hysterectomy and by £22 for an abdominal hysterectomy. Smoking is observed to reduce

Table 7.14   Potential binary covariates (%) by treatment and type of hysterectomy, EVALUATE

		Vaginal	Abdominal
Smoker	Laparoscopic	37.8	41.7
	Standard	40.3	47.1
Previous Pelvic Surgery	Laparoscopic	1.62	0.54
	Standard	0	0.73

Table 7.15   Parameter estimates, standard errors and *p*-values,
EVALUATE

Parameter		Vaginal		Abdominal	
		No covariates	Final model*	No covariates	Final model*
$\hat{\Delta}_c$	Estimate	400.8	425.2	185.8	177.7
	SE	69.79	70.72	100.8	102.8
	p-value	< 0.0001	< 0.0001	0.065	0.084
$\hat{\theta}_A$	Estimate		24.05		21.93
	SE	n.i.m.	5.399	n.i.m.	9.744
	p-value		< 0.0001		0.024
$\hat{\theta}_B$	Estimate				18.21
	SE	n.i.m.	n.i.m.	n.i.m.	8.169
	p-value				0.026
$\hat{\theta}_M$	Estimate		−185.5		
	SE	n.i.m.	73.700	n.i.m.	n.i.m.
	p-value		0.012		
$\hat{\theta}_P$	Estimate	.			2397
	SE	n.i.m.	n.i.m.	n.i.m.	1201
	p-value				0.046
$\hat{\Delta}_e$	Estimate	0.001542	0.003831	0.009148	0.005077
	SE	0.01031	0.009778	0.01008	0.009726
	p-value	0.881	0.695	0.364	0.602
$\hat{\omega}_Q$	Estimate		0.1505		0.1656
	SE	n.i.m.	0.02405	n.i.m.	0.01918
	p-value		< 0.0001		< 0.0001
$\hat{\omega}_M$	Estimate				−0.0242
	SE	n.i.m.	n.i.m.	n.i.m.	0.009110
	p-value				0.010
$\hat{\omega}_P$	Estimate				−0.1394
	SE	n.i.m.	n.i.m.	n.i.m.	0.06391
	p-value				0.029
$\hat{C}\left(\hat{\Delta}_e, \hat{\Delta}_c\right)$		−0.08368	−0.1034	−0.2285	−0.2411

*Includes only significant (*p* < 0.05) variables.
n.i.m. not in model.

Table 7.16 Unadjusted and adjusted ICER and INB with 90% confidence limits, EVALUATE

		Unadjusted	Adjusted
Vaginal	ICER	259 983	111 005
	(confidence limits)	(19 981, −25 910)	(19 502, −35 144)
	INB(10 000)	−385	−387
	(confidence limits)	(−600, −169)	(−599, −175)
	INB(20 000)	−369	−348
	(confidence limits)	(−740, 0.325)	(−706, 9.47)
Abdominal	ICER	20 312	34 998
	(confidence limits)	(1438, −25 649)	(900, −14 404)
	INB(10 000)	−94.3	−127
	(confidence limits)	(−353, 165)	(−386, 132)
	INB(20 000)	−2.86	−76.2
	(confidence limits)	(−405, 400)	(−473, 320)

cost by £186 for a vaginal hysterectomy. Not surprisingly, higher EQ-5D measures at randomization are associated with higher QALYs during follow-up for either type of hysterectomy. For abdominal hysterectomy, a one-unit increase in BMI is observed to increase cost by £18. Also for abdominal hysterectomy, previous surgery is observed to increase cost by £2397 and reduce QALYs in follow-up by 0.1394. When the significant covariates are adjusted for, the treatment difference in effectiveness more than doubles for vaginal hysterectomies and almost halves for abdominal hysterectomies. These differences have a substantial effect on the ICER, demonstrating the instability of the ratio statistics, but almost no effect on the INB evaluated at either £10 000 or £20 000 per QALY, see Table 7.16.

The only significant covariate by treatment interaction was with smoking in the cost model for the vaginal group. The coefficient was −505.30, with a p-value of 0.0003. The negative coefficient indicates that the increase in cost associated with the laparoscopic procedure was observed to be £505.30 more for non-smokers (£624.35 versus £119.05). To examine the interaction between treatment and smoking on the INB scale, a model was fitted that added the smoking by

Table 7.17   INB interaction between treatment and smoking, EVALUATE, vaginal

λ	$\hat{\omega}_{T \times M} \lambda - \hat{\theta}_{T \times M}$	S.E.	p-value
5000	182.4	65.61	0.0028
10 000	583.1	258.3	0.0239
15 000	621.9	346.3	0.0724
20 000	660.8	439.2	0.1323
25 000	699.7	534.4	0.1904

treatment interaction to the cost model, and smoking and the smoking by treatment interaction to the effectiveness model, to the final model given in Table 7.15. Then the hypothesis H : $\omega_{T \times M} \lambda - \theta_{T \times M} = 0$ was tested, where $T \times K$ indicates the treatment by smoking interaction. The estimate $\hat{\omega}_{T \times M} \lambda - \hat{\theta}_{T \times M}$, its standard error and associated p-value are given in Table 7.17 for several values of λ. For all values of λ the estimate is positive, indicating that the INB is observed to be greater among smokers, reaching significance for WTP values less than £10 000/QALY. Since there was no *a priori* reason to suspect an interaction of this type, and because several possible interactions were explored, these results should be considered exploratory at best.

7.4   SUMMARY

In this chapter, regression methods for the covariate adjustment of cost-effectiveness analysis were reviewed. Early work in this area has employed the net benefit directly in a regression framework (see for example Hoch, Briggs and Willan, 2002). However, further work has identified multivariable regression, in particular methods of seemingly unrelated regression, as a promising approach given that it allows different covariates to impact on cost and effectiveness, in which case efficiency gains can be realized.

Regression methods allow covariate adjustment at two levels. First, by using covariates to explain any chance imbalances between arms in a trial, an adjusted estimate of treatment effect can be presented.

The importance of adjustment of QALY outcomes for any baseline imbalance in quality of life measures has recently been highlighted (Manca *et al.*, 2005a). Even where there is no appreciable imbalance, adjustment for covariates can lead to an increased precision in the estimation of treatment effect by explaining some of the between-patient variability (Altman, 1985; Pocock, 1984). Second, adjusting for covariates allows for the examination of interactions that can identify sub-groups in which cost-effectiveness differs. While analysts should be cautious in their approach to sub-group analysis to avoid the potential pitfalls (Oxman and Guyatt, 1992; Collins *et al.*, 1987) it is often the case that the potential for heterogeneity in cost-effectiveness by patient characteristics is a prime concern of decision makers.

8

Multicenter and Multinational Trials

8.1 INTRODUCTION

Most clinical trials performed today are multicenter and many are multinational. The inclusion of multiple centers and countries affords greater statistical power resulting from an increase in sample size. Other advantages include the perception of greater generalizability and the opportunity for the sponsor, in the case of a drug trial, to use the results for registration in more than one country. There are a number of challenges associated with undertaking cost-effectiveness analysis using data from multinational clinical trials. The key issue relates to whether the data required for cost-effectiveness analysis can be pooled across the different countries. Most analyses of such trials fail to account for the multilevel nature of the sampling and will underestimate the uncertainty in the estimation of treatment effects if an interaction exists between center or country and treatment, i.e. if the treatment effects differ between centers or countries.

For the remainder of this section when we use the term unit we are referring to either center or country. Most analysis of multi-unit trials ignore unit effects by assuming (for a single outcome) that

$$y_{ki} = \theta + \delta t_{ki} + e_{ki} \tag{8.1}$$

where y_{ki} is the outcome for patient i from unit k; $t_{ki} = I\{\text{patient } i \text{ from unit } k \text{ is randomized to Treatment}\}$, and, e_{ki} is a random variable

Statistical Analysis of Cost-effectiveness Data. A. Willan and A. Briggs
© 2006 John Wiley & Sons, Ltd.

with mean zero. For this model the outcome has mean θ for patients on Standard and mean $\theta + \delta$ for patients on Treatment, with a treatment effect of δ. The assumption that these parameters are the same for all units is highly questionable, given that patient referral patterns, recruitment procedures, care and outcome assessment are likely to vary between units, especially between countries. If a treatment by unit interaction exists, i.e. if δ varies by unit, the above model will underestimate the variance in the estimation of δ, leading to inflated type I error rates. In the case of multinational trials some authors (Willke *et al.*, 1998; Koopmanschap *et al.*, 2001; Cook *et al.*, 2003), have proposed a fixed-effect model given by

$$y_{ki} = \theta_k + \delta_k t_{ki} + e_{ki} \qquad (8.2)$$

where θ_k and δ_k are fixed parameters. For this model the mean outcome in each treatment group and the treatment effect vary between units. However, this model provides no estimate of overall treatment effect, which is particularly problematic for multicenter trials in a single country where one might want to make inference for the country as a whole. In addition, the estimates of the treatment effect for an individual unit are based on that unit's data only. This leads to very inefficient estimation and is particularly problematic for multinational trials where estimates of treatment effects for individual countries are usually required.

As an alternative many authors (Nixon and Thompson, 2005; Pinto *et al.*, 2005; Manca *et al.*, 2005b; Willan, Pinto, O'Brien *et al.*, 2005; Grieve *et al.*, 2005; Brown and Kempton, 1994; Skene and Wakefield, 1990) have proposed random effects models characterized by

$$y_{ki} = \theta_k + \delta_k t_{ki} + e_{ki} \qquad (8.3)$$

where θ_k and δ_k are assumed to be normally distributed with respective means θ and δ and variances σ_θ^2 and σ_δ^2. The overall treatment effect is given by δ, while the treatment effect for unit k is given by δ_k, and is estimated using data from all units, as illustrated in subsequent sections. There are two analytic approaches to parameter estimation for these random effect models. The first, which we will call aggregate level analysis (ALA), is appropriate when data are presented to the analyst in aggregate form, i.e. as estimated means and variances by unit. Data are often presented in this way to protect patient anonymity. The second method, which we will refer to as hierarchical modeling (HM), is appropriate when individual patient data are available. When patient level data are available HM is preferred to ALA for several reasons. First, it allows for the use of more flexible distributions for modeling cost data. Cost data are almost always skewed to the right, being bounded on the left by zero, and although the sample mean often provides a reasonable estimate, it is generally inefficient, and more efficient estimation may be provided by fitting a gamma or lognormal distribution. Second, HM allows for covariate adjustment for patient level variables. Third, most multinational trials are also multicentered, and HM allows for the modeling of multiple sources of error, i.e. patients within center, centers within countries. Fourth, HM provides for a more appropriate adjustment in the estimation of variance components, whereas ALA adjustments for variance component estimation is somewhat *ad hoc*.

This chapter is divided into three sections. The first offers some background to the issue of multinational clinical trials, including potential threats to the transferability of data between countries and a taxonomy of different studies based on the level of pooling undertaken. The second reviews the fixed effect approaches to multinational analysis. The third looks at random effects approaches including both aggregate and individual patient data level analysis.

8.2 BACKGROUND TO MULTINATIONAL COST-EFFECTIVENESS

In an early contribution to the literature, O'Brien (1997) identified six threats to the transferability of data for economic analysis for questions of whether treatments that are cost-effective in one country might not necessarily be cost-effective in another. These six threats are also useful to consider in the context of multinational clinical trials.

Demography and epidemiology of disease. The underlying premise of a multinational clinical trial is that the treatment effect on the underlying biological process is constant across countries. However, difference in demography and epidemiology of disease between countries may threaten this assumption, particularly with respect to the *absolute* benefit of treatment in different countries. The treatment effect in most clinical trials is often a relative measure, such as hazards ratios, relative risks or odds ratios. In such cases, it may be appropriate to assume constant relative treatment effects across countries, even when baseline epidemiology/demography differs (McAlister, 2002).

Clinical practice and conventions. In a clinical trial, of course, the differences in clinical practice and conventions between countries are limited by the protocol of the trial. Nevertheless, there is the potential for differences in provision of 'usual care' (as opposed to the treatments under evaluation) to impact the comparison between countries, For example, it is well known that in the US, rates of surgical intervention compared to medical intervention, tend to be higher than in many other health systems.

Incentives and regulations for health care providers. Different countries will have different incentives and regulations which will result in practice variations between countries. These in turn may result in different levels of resource use across different categories of care.

Relative price levels. Absolute price levels clearly differ between countries, but it is the relative difference between different categories of resource use between countries that is potentially problematic for multinational trials. Economic theory suggests that differences in relative prices will result in substitution from relatively more expensive resources to relatively cheaper ones. Therefore, differences in relative prices between countries could lead to different practice patterns.

Consumer preferences. Quality of life measures, that are used in the calculation of QALYs are based upon individual preferences. There is no reason to suppose that these preferences are not culturally dependent such that we might expect differences to be observed between countries.

Opportunity costs of resources. Different countries will have different levels of ability to pay for improved health outcomes. What is considered cost-effective in a health system in North America or Western Europe, may not be considered affordable in South America or Eastern Europe. This fundamentally limits the usefulness of an overall cost-effectiveness analysis of the trial itself.

Each of these six threats to transferability of economic data give reasons why we might be concerned with making a single estimate of cost-effectiveness across all countries. Yet, as was argued in the introduction, the rationale for multinational clinical trials is usually related to obtaining a large study with increased power to detect treatment effects. The fundamental problem, therefore, relates to whether data are pooled to maximize power or split to maximize the credibility of the economic analysis in each individual country.

As part of a workshop to explore whether it was possible to gain consensus on how to analyse multinational clinical trials for economic appraisal, a taxonomy of different approaches to the analysis were offered (Reed *et al.*, 2005). This taxonomy is presented in Table 8.1, together with the categorization of 18 multinational studies reviewed in the cardiology field. The categorization relates to the intersection between three factors: whether the measure of clinical effectiveness data was pooled across all countries or split into each country; whether the measure of resource use data was obtained by pooling across all countries or by splitting out each country's resource use; and whether unit costs were presented for a single country, or from a weighted average of all countries. Fully pooled analyses were defined as those for which both resource use and effectiveness was estimated from a pooled analysis of the all of the data. Fully split analyses were defined as those studies that looked only within the individual country for estimates of cost and effectiveness. Partially split analyses were defined where effectiveness was obtained from a pooled estimate, but resource use was split out by individual country (although the reverse could also true this was not considered in the paper, presumably as this was felt to be an unlikely approach in practice). For each of these three broad categories, sub-categories were defined by the approach to costing as described above.

What is clear from the review of 18 cardiology trials (even given the small sample size) is that the fully pooled approach has been the most prevalent approach to date, with half of all the studies presenting their analysis in this way. Fully split analyses are much less common, with only two studies presenting this approach. Only two studies presented a 'multicountry' approach to costing where results were aggregated across all participating centres. Much more common was to present the results from the perspective of a single country, which reflects the importance of the decision makers being located in individual countries.

Table 8.1 Terminology for methodological approaches to multinational economic evaluations (adapted from *Reed et al.*, 2005)

Source of clinical effectiveness data	Source of resource use data	Type of analysis		Costing methodology		Classification	Categorization of 18 studies from the cardiology field
All participating countries	All countries	Fully pooled	→	Multicountry	→	Fully pooled with multicountry costing	2
			→	One country	→	Fully pooled with one-country costing	9
All participating countries	One country or sub-set of countries	Partially split	→	Multicountry*	→	Partially split with multicountry costing	0
			→	One country	→	Partially split with one-country costing	5
One country or sub-set of countries	One country or sub-set of countries	Fully split	→	Multicountry*	→	Fully split with multicountry costing	0
			→	One country	→	Fully split with one-country costing	2

* Multicountry costing is feasible only if resource use data are available from more than one country.

8.3   FIXED EFFECT APPROACHES

8.3.1   Willke *et al.*

One of the earliest attempts to address the statistical analysis of multinational clinical trials for a cost-effectiveness analysis was presented by Willke *et al.* (1998). They examined how cost and effectiveness interact using data from a multinational clinical trial of treatment of subarachnoid haemorrhage. Using a series of regression analyses, they developed a novel approach that explored the treatment by country interactions in both effectiveness and cost and which allowed the treatment effect on cost to be estimated independently of the treatment effect on effectiveness.

They presented results for five separate countries (not identified) corresponding to the most common categories from Table 8.1. These results are presented in Figure 8.1, with the different country estimates represented simply by the numerals 1–5. These results clearly emphasize the differing spread of estimates that is obtained under the

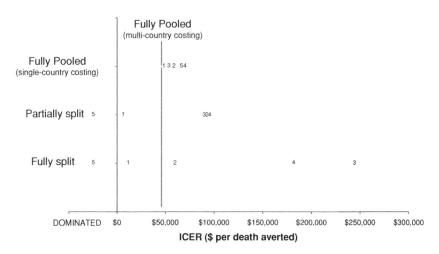

Figure 8.1 Alternative approaches to estimating country-specific cost-effectiveness estimates from a multinational trail of subarachnoid haemorrhage

different approaches. A fully pooled analysis with multicountry costing produces just a single estimate for the whole trial (represented by the vertical line in the figure). A fully pooled analysis with price weights from the individual countries produces very little variability in the results. Pooling the effectiveness while splitting out the cost (with countries own price weights) provides a much greater spread, but the widest variation comes from the fully split analysis.

8.3.2   Cook *et al.*

The increasing spread of results as the data are more widely split that is shown in Figure 8.1 is entirely consistent with expectations. The smaller sample sizes involved in the split analyses will increase variability. The key question is to what extent is this variability related to random error or to what extent does it reflect systematic differences in the cost-effectiveness between countries due to the sorts of factors discussed in Section 8.2?

In a more recent contribution, Cook *et al.* (2003) proposed the use of standard tests of homogeneity in treatment effect (Gail and Simon, 1985), to inform the decision of whether or not it was appropriate to pool cost data across countries. They outline methods based both on the angular transformation of the ICER and INB using the 4S study of cholesterol reduction with simvastatin as an example (Jonsson *et al.*, 1996; Scandinavian Simvastatin Survival Study Group, 1994) . The results of their INB analysis for the countries of Denmark, Finland, Iceland, Norway and Sweden are presented in Figure 8.2 for a willingness-to-pay threshold of $75 000 per additional survivor.

The results show that there is some variability when country-specific subsets are analysed. Positive net benefit is observed for Denmark, Norway and Sweden, while negative net benefits are observed in Norway and Iceland. Nevertheless all of the confidence limits overlap zero, and tests for both quantitative and qualitative interaction are not significant. The authors suggest that in the absence of strong evidence of heterogeneity, it is appropriate to consider pooling these data, and the overall pooled estimate (ignoring country) is

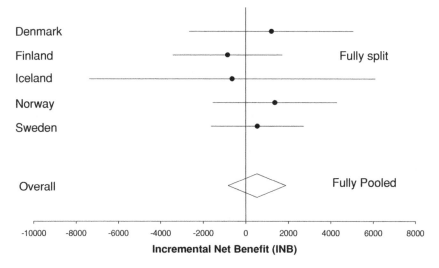

Figure 8.2 Incremental net benefit of simvastatin treatment in the 4S study overall and by country assuming a threshold willingness-to-pay of $75 000 per additional survivor. Tests for both quantitative and qualitative country by treatment interactions are insignificant. Adapted from Cook *et al.* (2003)

clearly much more precise with a much tighter confidence interval, which nevertheless still overlaps zero net benefit.

The authors are careful to point out that these tests often suffer from low power. This is perhaps unsurprising given that part of the rationale for multinational trials is to achieve sufficient power *overall* on effectiveness. The authors suggest that evidence of a country-by-treatment interaction is likely to provide an argument against pooling the data, but that absence of evidence should not necessarily be interpreted as a rationale to pool. Although no partially split analysis is presented, there would clearly be an opportunity to employ the approach outlined by Cook *et al.* to provide the rationale for such an analysis if, for example, no convincing evidence of a country-by-treatment interactions were found for the effectiveness data, but evidence of heterogeneity was found for cost. Given the relative similarity of the Scandinavian countries and their health systems, the lack of heterogeneity in this case is not unexpected. For multinational trials covering a broader range of countries, evidence of heterogeneity is more likely.

8.4   RANDOM EFFECTS APPROACHES

The potential problem with the fixed effect approaches identified in Section 8.3 is that they require a choice to be made between pooling or splitting. Although when splitting the data, random error is important, systematic differences between countries are also likely to be important. Random effects models offer the potential to estimate systematic differences between countries, while simultaneously adjusting for their expected random error associated with splitting the data. In this regard, they offer something of a statistical middle ground between a fully split analysis and a fully pooled analysis. In Sections 8.4.1 and 8.4.2 a random effects analysis based on aggregate measures is outlined for multicenter and multinational trials, respectively, while in Section 8.4.3 the use of hierarchical modeling when patient level information is available is described.

8.4.1   Aggregate level analysis: multicenter trials

Consider a multicenter trial carried out in a single country. Assuming that the centres in the trial are reasonably representative and that inference is to be made for (and health policy applied to) the country as a whole, then the most straightforward analysis is analogous to a random effects meta-analysis where, instead of summarizing over trials, the analyst summarizes over centers.

Suppose there are M centers, and let $\Delta_k = (\Delta_{ek}, \Delta_{ck})^T$ be the vector containing the expected difference in effectiveness and cost in center k. Let $\hat{\Delta}_k$ be the estimate of Δ_k based on the data from center k using the appropriate procedures discussed in previous chapters, and let $V_k = V(\hat{\Delta}_k | \Delta_k)$. Furthermore, assume that $\hat{\Delta}_k | (\Delta_k, V_k) \sim N(\Delta_k, V_k)$ and $\Delta_k | (\Delta, U) \sim N(\Delta, U)$. The pooled overall estimator of Δ is given by $\hat{\Delta} = \hat{W}^{*-1} \sum_k \hat{W}_k^* \hat{\Delta}_k$, where $\hat{W}_k^* = (\hat{V}_k + \hat{U})^{-1}$, $\hat{W}^* = \sum_k \hat{W}_k^*$, $\hat{V}_k$ is the estimate of V_k, based on the data from centre

k, and $\hat{U}$ is an estimate of U. All sums are from $k = 1$ to M. The estimator of $V(\hat{\Delta})$ is given by $\hat{V}(\hat{\Delta}) = \hat{W}^{*-1}$. The quantities $\hat{\Delta}$ and $\hat{V}(\hat{\Delta})$ provide the five parameter estimates that are required to perform an overall cost-effectiveness analysis as discussed in Chapter 4.

There are three possible estimators for U. The first, referred to as unweighted, is given by

$$\hat{U}_{UW} = \frac{\sum_k \left(\hat{\Delta}_k - \hat{\Delta}_{UW}\right) \left(\hat{\Delta}_k - \hat{\Delta}_{UW}\right)^T - \left(1 - M^{-1}\right) \sum_k \hat{V}_k}{M - 1}$$

where $\hat{\Delta}_{UW} = \sum_k \hat{\Delta}_k / M$. $\hat{U}_{UW}$ is unbiased. The second estimator of U, referred to as weighted, is given by

$$\hat{U}_{W\{ij\}} = \frac{\left[\sum_k \hat{W}_k^{1/2} \left(\hat{\Delta}_k - \hat{\Delta}_W\right) \left(\hat{\Delta}_k - \hat{\Delta}_W\right)^T \hat{W}_k^{1/2}\right]_{\{ij\}} - c_{ij}}{m_{ij}}$$

where the $A_{\{ij\}}$ is the ith–jth element of the matrix A, $\hat{W}_k$ is a diagonal matrix, with kth diagonal entry given by

$$\hat{W}_{k\{ii\}} = \hat{w}_{ki} = \frac{\left(\hat{V}_{k\{ii\}}\right)^{-1}}{\sum_j \left(\hat{V}_{j\{ii\}}\right)^{-1}}$$

$$\hat{\Delta}_W = \sum_k \hat{W}_k \hat{\Delta}_k$$

$$c_{ij} = \sum_k \hat{w}_{ki}^{1/2} \hat{w}_{kj}^{1/2} \hat{V}_{k\{ij\}} + \sum_k \hat{w}_{ki}^{1/2} \hat{w}_{kj}^{1/2} \sum_l \hat{w}_{li} \hat{w}_{lj} \hat{V}_{l\{ij\}}$$
$$- \sum_k \hat{w}_{ki}^{3/2} \hat{w}_{kj}^{1/2} \hat{V}_{k\{ij\}} - \sum_k \hat{w}_{ki}^{1/2} \hat{w}_{kj}^{3/2} \hat{V}_{k\{ij\}}$$

and

$$m_{ij} = \sum_k \hat{w}_{ki}^{1/2} \hat{w}_{kj}^{1/2} + \sum_k \hat{w}_{ki}^{1/2} \hat{w}_{kj}^{1/2} \sum_l \hat{w}_{li} \hat{w}_{lj}$$
$$- \sum_k \hat{w}_{ki}^{3/2} \hat{w}_{kj}^{1/2} - \sum_k \hat{w}_{ki}^{1/2} \hat{w}_{kj}^{3/2}$$

$\hat{U}_W$ is also unbiased. The diagonal elements of $\hat{U}_W$ are equal to the estimator of between-country variance based on the method of moments for a random effects model (see Dersimonian and Laird, 1986).

A third estimator of U, denoted $\hat{U}_{\text{ReML}}$, is the restricted maximum likelihood estimator of U, achieved through a combination of the EM algorithm, with an Aitken acceleration, and the Newton–Raphson algorithm (see Pinto *et al.*, 2005). Except for $\hat{U}_{\text{ReML}}$, which is maximized over the space of non-negative definite matrices, the diagonal elements of the estimate of U can be negative. This implies that the observed between-center variance of the estimator of the between-treatment difference in effectiveness or cost is less than would be expected, given the observed variance among patients within a center. Under such circumstances the negative diagonal element and the corresponding off-diagonal elements should be set to zero.

8.4.2 Aggregate level analysis: multinational trials

For multinational trials let Δ_k be the vector containing the expected difference in effectiveness and cost in country k. Also, let $\hat{\Delta}_k$ be the estimate of Δ_k and $\hat{V}_k$ be the estimate of $V\left(\hat{\Delta}_k\right)$ based on the data from country k using the appropriate procedures discussed in previous chapters and Section 8.1.1. Using the procedures outlined in Section 8.1.1, a pooled overall estimate of Δ and its corresponding variance can be calculated. These would be the appropriate estimates for inference about a world for which the countries in the trial are reasonably representative. However, no one makes inference, let alone health policy, on an multinational basis, and these estimates are not appropriate for any particular jurisdiction. What health policy makers need to know is what inference can be made for their country based on the results of the trial? Individual country inference can be made using shrinkage estimators (see Pinto *et al.*, 2005; Willan, Pinto, O'Brien *et al.*, 2005; Efron and Morris, 1972).

In the empirical Bayes framework the prior density for Δ_k is $N(\Delta, U)$ and the posterior density, given the data from country k, is $N(\tilde{\Delta}_k, (V_k^{-1} + U^{-1})^{-1})$, where

$$\tilde{\Delta}_k = \left(V_k^{-1} + U^{-1}\right)^{-1}\left(V_k^{-1}\hat{\Delta}_k + U^{-1}\Delta\right)$$
$$= (V_k + U)^{-1}\left(U\hat{\Delta}_k + V_k\Delta\right) = \Delta_k + V_k\left(U + V_k\right)^{-1}(\Delta - \Delta_k)$$

An estimator of $\tilde{\Delta}_k$ is given by

$$\hat{\tilde{\Delta}}_k = \hat{\Delta}_k + \frac{M-4}{M-1} \hat{V}_k \left(\hat{U} + \hat{V}_k\right)^{-1} \left(\hat{\Delta} - \hat{\Delta}_k\right)$$

and is known as the shrinkage estimator for Δ_k because, as U approaches 0, it approaches (i.e. shrunken towards) the overall estimator $\hat{\Delta}$. The factor $(M-4)/(M-1)$ is for bias correction to account for variance estimation. An estimator for the variance of $\hat{\tilde{\Delta}}_k$ is given by

$$\hat{V}\left(\hat{\tilde{\Delta}}_k\right) = \hat{V}_k - \frac{M-4}{M-1} \hat{V}_k \left(\hat{U} + \hat{V}_k\right)^{-1} \hat{V}_k + P_k \hat{W}^{*-1} P_k^T$$

$$+ \operatorname{diag}\left\{v_k \times \left(\hat{\Delta}_k - \hat{\Delta}\right) \times \left(\hat{\Delta}_k - \hat{\Delta}\right)\right\}$$

where $P_k = (M-4)/(M-1)\hat{V}_k \left(\hat{U} + \hat{V}_k\right)^{-1}$, $\times$ denotes componentwise multiplication of vectors, v_k is a vector of length 2 with ith component equal to

$$\frac{2(M-4)^2 \left(\hat{V}_{k\{ii\}}\right)^2 \left(\bar{v}_i + \hat{U}_{\{ii\}}\right)}{(M-1)^2 (M-3) \left(\hat{V}_{k\{ii\}} + \hat{U}_{\{ii\}}\right)^3}$$

and $\bar{v}_i$ depends on the estimator used for U. For $\hat{U}_{UW}$, $\bar{v}_i = \Sigma_k V_{k\{ii\}}/M$, for $\hat{U}_W$, $\bar{v}_i = M(\Sigma_k(\hat{V}_{k\{ii\}})^{-1})^{-1}$, and for $\hat{U}_{ReML}$, $\bar{v}_i = \{\Sigma_k(\hat{W}^*_{k\{ii\}})^2\}^{-1}\Sigma_k(\hat{W}^*_{k\{ii\}})^2 \hat{V}_{k\{ii\}}$. For more detail, the reader is referred to Pinto *et al.* (2005).

The shrinkage estimators are based on the assumption that the between-country variation in treatment differences is normally distributed. However, Pinto *et al.* (2005) demonstrate that they perform well, even when the between-country variation is uniform. Nonetheless, methods proposed by Hardy and Thompson (1998) can be used to test the normality assumption.

8.4.2.1 Aggregate level analysis: example: the ASSENT-3 trial

The Assessment of the Safety and Efficacy of New Thrombolytic Regimens (ASSENT-3) trial has been described elsewhere in detail (see

The ASSENT-3 Investigators, 2001). Patients with ST elevation acute myocardial infarction were randomized between three treatment arms:

- heparin: full-dose tenecteplase plus unfractionated heparin;
- enoxaparin: full-dose tenecteplase plus enoxaparin;
- abciximab: half-dose tenecteplase plus unfractionated heparin plus abciximab.

The enoxaparin and abciximab arms were compared to the heparin arm with respect to safety and effectiveness. A total of 6095 patients were enrolled at 575 clinical sites in 26 countries, with the largest number of patients (975) from the US and the smallest number (23) from Finland. A total of 435 patients were enrolled from Canada. The measure of effectiveness was freedom from death, in-hospital re-infarction and refractory ischemia for 30 days. Both investigational regimens (the enoxaparin and abciximab arms) showed a statistically significant improvement in effectivenvess compared with the heparin arm. Success (i.e. freedom from death, in-hospital re-infarction and re-fractory ischemia) was observed in 1721 of the 2036 (84.5%) patients randomized to the heparin arm, versus 1804 of the 2037 (88.6%) patients randomized to the enoxaparin arm and 1794 of the 2017 (88.9%) patients randomized to the abciximab arm.

As part of the trial, resource use data were collected on length of hospital stay, study drug and dosage, cardiac catheterization, percutaneous coronary intervention (PCI) and coronary artery bypass graft (CABG) surgery during the index hospitalization and length of stay and PCI and CABG surgery use during repeat hospitalizations within 30 days. Canadian price weights (e.g. drug prices) for resources consumed were used to provide total 30-day cost for each patient.

Estimates for U, Δ and $V(\hat{\Delta})$ for the various methods are shown in Table 8.2 for the enoxaparin versus heparin (E–H) and abciximab versus heparin (A–H) comparisons. For E–H the elements of $\hat{U}$ for weighted and unweighted are 0. Therefore $\hat{\Delta}$ will be the same for weighted and unweighted and is equivalent to the fixed effect estimate from a meta-analysis where the analyst is pooling over

Table 8.2 Overall estimation for the ASSENT-3 trial

	$\hat{U}$	$\hat{\Delta}$	$\hat{V}(\hat{\Delta})$
Enoxaparin vs heparin			
Unweighted	$\begin{pmatrix} 0 & 0 \\ 0 & 0 \end{pmatrix}$	$\begin{pmatrix} 0.03204 \\ -57.02 \end{pmatrix}$	$\begin{pmatrix} 0.00009917 & -0.6113 \\ -0.6113 & 18757 \end{pmatrix}$
Weighted	$\begin{pmatrix} 0 & 0 \\ 0 & 0 \end{pmatrix}$	$\begin{pmatrix} 0.03204 \\ -57.02 \end{pmatrix}$	$\begin{pmatrix} 0.00009917 & -0.6113 \\ -0.6113 & 18757 \end{pmatrix}$
ReML	$\begin{pmatrix} 0.0002417 & 1.013 \\ 1.013 & 5236 \end{pmatrix}$	$\begin{pmatrix} 0.03381 \\ -52.97 \end{pmatrix}$	$\begin{pmatrix} 0.0001151 & -0.5501 \\ -0.5501 & 19255 \end{pmatrix}$
Abciximab vs heparin			
Unweighted	$\begin{pmatrix} 0.001801 & -6.983 \\ -6.983 & 401681 \end{pmatrix}$	$\begin{pmatrix} 0.03928 \\ 921.5 \end{pmatrix}$	$\begin{pmatrix} 0.0001994 & -1.058 \\ -1.058 & 41345 \end{pmatrix}$
Weighted	$\begin{pmatrix} 0.0003038 & -0.005418 \\ -0.005418 & 135194 \end{pmatrix}$	$\begin{pmatrix} 0.03741 \\ 942.7 \end{pmatrix}$	$\begin{pmatrix} 0.0001198 & -0.6627 \\ -0.6627 & 27269 \end{pmatrix}$
ReML	$\begin{pmatrix} 0.0002183 & 3.519 \\ 3.519 & 70033 \end{pmatrix}$	$\begin{pmatrix} 0.03730 \\ 948.7 \end{pmatrix}$	$\begin{pmatrix} 0.0001124 & -0.4410 \\ -0.4410 & 23375 \end{pmatrix}$

countries rather than studies. For this situation the shrinkage estimators simplify to

$$\frac{3}{M-1}\hat{\Delta}_k + \frac{M-4}{M-1}\hat{\Delta}$$

The diagonal elements of the ReML estimate of U are not negative, since they are constrained not to be. However, the ReML estimate of Δ, which in this case is a random effects estimate, is very similar to those for weighted/unweighted, but because it is a random effects estimate, its variance components will be larger. For the A–H comparison the elements of $\hat{U}$ are non-zero for all methods. The diagonal elements are smallest for ReML and largest for unweighted. The estimates of Δ are all very similar, and no consistent or material difference is seen between methods. The diagonal elements of $\hat{V}(\hat{\Delta})$ are smallest for ReML and largest for unweighted, which is consistent with our observations regarding $\hat{U}$.

Parameter estimates for Canada are shown in Table 8.3. The pooled estimates, using ReML, are given in the top panel. These, for obvious reasons, will have the smallest standard errors. However, as seen in the simulation results given by Pinto *et al.* (2005), confidence intervals based on these estimates and standard errors have very poor coverage for individual country parameters when $U \neq 0$. Estimates based on the Canadian data alone (i.e. those appropriate for a fixed effect model as in Model 8.2) are shown in the second panel. These, for obvious reasons, will have the largest standard errors, although the corresponding confidence intervals have the appropriate coverage. The shrinkage estimates are shown in the bottom three panels, for each method of estimating U. For both treatment comparisons and for both effectiveness and cost the 'weighted' estimates have smaller standard errors than the 'unweighted', and except for the effectiveness for the E–H comparison, the ReML estimates have smaller standard errors than 'weighted'.

The Canadian cost-effectiveness acceptability curves for the abciximab versus heparin contrast are shown in Figure 8.3. The curve based on the Canadian data alone is the least steep, and therefore would provide the widest credible intervals for the ICER. The curves for the shrinkage estimates show a consistent pattern, with ReML providing the narrowest intervals, followed by weighted and then

Table 8.3 Treatment contrasts for Canada

Estimate (std. error) [p-value] {95% C.I.*}	Enoxaparin vs heparin		Abciximab vs heparin	
	Effectiveness	Cost	Effectiveness	Cost
Pooled random effects (ReML)	0.03381 (0.01073) [0.001626] {0.01278, 0.05483}	−52.97 (138.8) [0.7026] {−324.9, 219.0}	0.03730 (0.01060) [0.0004352] {0.01652, 0.05808}	948.7 (152.9) [<0.0001] {649.0, 1248}
Canadian patients only	0.008773 (0.04067) [0.8292] {−0.07093, 0.08848}	−731.2 (615.1) [0.2346] {−1937, 474.5}	0.06032 (0.0374) [0.1068] {−0.01299, 0.1336}	1008 (663.9) [0.1289] {−293.1, 2309}
Shrinkage estimates				
Unweighted	0.02913 (0.02019) [0.1490] {−0.01043, 0.06870}	−141.3 (392.8) [0.7191] {−911.1, 628.6}	0.05231 (0.03008) [0.08205] {−0.006651, 0.1113}	962.5 (497.6) [0.05304] {−12.65, 1938}
Weighted	0.02913 (0.01856) [0.1166] {−0.007253, 0.06552}	−141.3 (324.0) [0.6627] {−776.2, 493.7}	0.04408 (0.02231) [0.04822] {0.0003436, 0.08782}	981.2 (396.9) [0.01343] {203.3, 1759}
ReML	0.02512 (0.02210) [0.2557] {−0.01820, 0.06845}	−162.7 (302.6) [0.5907] {−755.8, 430.4}	0.04350 (0.02039) [0.03288] {0.003539, 0.08345}	1012 (339.2) [0.002853] {347.0, 1676}

* Credible intervals.

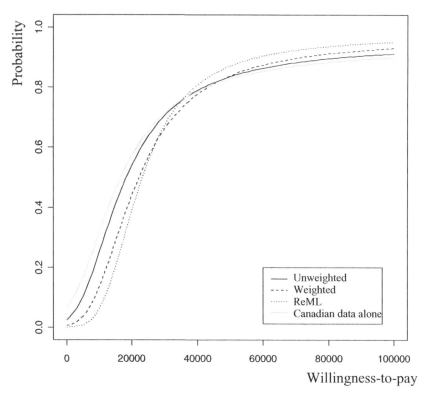

Figure 8.3 Canadian cost-effective acceptability curves for abciximab versus heparin

unweighted. The curves for the enoxaparin versus heparin contrast are shown in Figure 8.4. The shrinkage estimates show no consistent pattern. The curve based on the Canadian data alone has negative slope because of the large amount of probability in the SW quadrant, i.e. where enoxaparin is less costly and less effective.

8.4.3   Hierarchical modeling

Hierarchical modeling (HM) provides a more flexible approach for addressing issues posed by multinational and multi-centre trials (see Nixon and Thompson, 2005; Manca *et al.*, 2005b; Grieve *et al.*, 2005; Brown and Kempton, 1994; Skene and Wakefield, 1990). Most

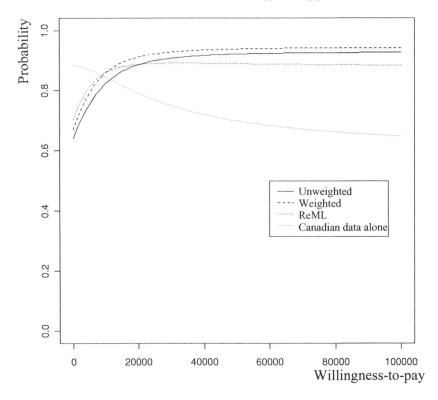

Figure 8.4 Canadian cost-effective acceptability curves for enoxaparin versus heparin

importantly, HM allows for distributions other than the normal to be used for modeling cost data, which are often highly skewed to the right. Furthermore, the addition of patient-level and country-level covariates is relatively straightforward. Adopting the HM formulation as first proposed by Nixon and Thompson (2005) we have

$$E_{ji} \sim \text{Dist} \left(\phi_{Eji}, \sigma_j \right) \qquad \text{and } C_{ji} \sim \text{Dist} \left(\phi_{Cji}, \omega_j \right)$$

where, recalling from Section 6.1, E_{ji} and C_{ji} are the observed effectiveness and cost, respectively, for patient i on arm j ($j = T, S$). Thus E_{ji} and C_{ji} are distributed with some (at this point) unspecified distribution with mean and standard deviation given by ϕ_{Eji} and σ_j and ϕ_{Cji} and ω_j, respectively. In a multinational trial of M countries the HM is specified by

$$\phi_{Eji} = \mu_{Ej} + \sum_{k=1}^{M} x_{jik} \{\gamma_{Ek} + I\,(j = T)\,\delta_{Ek}\} + \beta_j \left(C_{ji} - \phi_{Cji} \right)$$

$$\phi_{Cji} = \mu_{Cj} + \sum_{k=1}^{M} x_{jik} \{\gamma_{Ck} + I\,(j = T)\,\delta_{Ck}\}$$

where $x_{jik} = I$(patient i on arm j is from country k) and $I(\cdot)$ is the indicator function. Non-zero values for β_j induce a correlation between cost and effects, which can be different for each arm. The quantities γ_{Ek}, δ_{Ek}, γ_{Ck} *and* δ_{Ck} are random variables with mean zero, which implies that country is a random effect and interacts with treatment arm. The overall treatment effects are $\Delta = (\mu_{ET} - \mu_{ES} + \tau_E,\ \mu_{CT} - \mu_{CS} + \tau_C)^T$ and the country-specific treatment effects are $\tilde{\Delta}_k = \Delta + (\delta_{Ek},\ \delta_{Ck})^T$, where $\tau_E = \sum_{k=1}^{M} p_k \delta_{Ek}$, $\tau_C = \sum_{k=1}^{M} p_k \delta_{Ck}$ and p_k is the proportion of the total number of patients that come from country k. The between-country variance U is given by posterior variance of $(\delta_{Ek},\ \delta_{Ck})^T$. Parameter estimation is facilitated by Markov chain Monte Carlo methods using the software BUGS (see Spiegelhalter, Thomas and Best, 1999).

8.5 SUMMARY

In this chapter we have outlined some of the issues and challenges facing analysts undertaking cost-effectiveness analyses using data from multinational and multicenter clinical trials. It is important to emphasize that these methods are in their infancy and there are few examples in the current literature. Overall there is a lack of consensus as to how to conduct economic evaluation appropriately in such trials (Reed *et al.*, 2005). Although the fixed effect models reviewed here provide one approach to evaluation, it is argued that the necessity to choose between pooling or splitting the data could limit their usefulness. Random effects models offer a middle ground between these extremes, and we therefore believe that they represent a promising approach, reflecting as it does, the naturally hierarchical nature of the data.

9

Modeling Cost-effectiveness

9.1 INTRODUCTION

So far in this book the focus has been on what might be considered 'standard' or 'conventional' approaches to cost-effectiveness analysis when patient level information on costs and effects is available from clinical trials. This typically involves the direct estimation of the five pieces of information (mean incremental costs, effects, their variance and covariance) required to formulate either the ICER or the net benefit statistics. In this chapter, a more general framework of statistical modeling will be proposed, based on modeling the separate components of the cost-effectiveness equation, in order to build indirect estimates of incremental cost and effectiveness. Such an approach may have some advantages when it comes to handling practical problems associated with real-life analyses such as missing data, variability in high-cost rare events and the need to extrapolate beyond the follow-up period of the trial. The methods suggested in this chapter are similar to the general approach of modeling cost-effectiveness information as part of a decision analytic model. However, we do not attempt to address the full scope of decision analytic modeling methods here, instead we refer the interested reader to a number of excellent overviews (Kuntz and Weinstein, 2001). Rather, in this chapter, we focus on the statistical modeling of a single data source (such as a clinical trial) for the purposes of calculating cost-effectiveness. The chapter starts by outlining

Statistical Analysis of Cost-effectiveness Data. A. Willan and A. Briggs
© 2006 John Wiley & Sons, Ltd.

a general framework and then moves on to a specific case study: an economic analysis of a stratified randomized controlled trial in the area of asthma.

9.2 A GENERAL FRAMEWORK FOR MODELING COST-EFFECTIVENESS RESULTS

In a seminal article, Weinstein and Stason (1977) introduced cost-effectiveness analysis to a medical audience. In doing so, they emphasised an incremental approach to estimating additional costs and health gains (in terms of QALYs) and went on to describe net health care costs as consisting of several components

$$\Delta C = \Delta C_{Rx} + \Delta C_{SE} - \Delta C_{Morb} + \Delta C_{\Delta LE}$$

including differences in the health care costs of treatment (ΔC_{Rx}), differences in the health care costs related to adverse side-effects associated with treatment (ΔC_{SE}), differences in the costs associated with treating the adverse consequences of the disease (ΔC_{Morb}), and a final component representing the increased costs associated with differential life expectancy ($\Delta C_{\Delta LE}$).

Similarly, they described the net health effectiveness in terms of QALYs as also being comprised of a number of components

$$\Delta E = \Delta E_{LE} + \Delta E_{Morb} - \Delta E_{SE}$$

relating to differences in life-expectancy due to treatment (ΔE_{LE}), differences in quality of life due to treatment (ΔE_{Morb}), and quality of life effects related to side-effects (ΔE_{SE}).

The advantage of separating out the components of the incremental cost and incremental effectiveness estimates relates to the ability to model the components differently. For example, the vast majority of health outcomes from clinical trials are relative measures, such as relative risk reductions, odds and hazard ratios. By contrast, treatment side-effects are likely to be additive in that all patients may be at risk. A recent paper by Sutton *et al.* (2005), building on an earlier paper by Glasziou and Irwig (1995), demonstrates the statistical estimation of health outcome in the presence of a relative treatment effect and an additive side-effect profile.

The same argument can be applied to the estimation of incremental cost. For example, differences in the health care costs of treatment are likely to be additive, similarly the costs of managing any side-effects of treatment. By contrast, the reduction in costs from morbidity and the additional costs associated with increased life expectancy may best be modeled as a relative reduction, especially when these correspond to a reduction in serious adverse disease events, the risk of which may be reduced proportionally by treatment.

The statistical modeling of the components of the cost-effectiveness calculus can lead to an improved understanding of how the treatments, and other important prognostic factors, interact to form cost-effectiveness. This can be important for specifying sub-group effects which do not suffer from the well-known problems of spurious results due to data splitting and can help 'demystify' cost-effectiveness for clinical colleagues who are concerned with the way in which the estimates are formed. Nevertheless, as with any modeling, the validity of the resulting estimates will only be as credible as the underlying modeling assumptions.

9.3   CASE STUDY: AN ECONOMIC APPRAISAL OF THE GOAL STUDY

The Gaining Optimal Asthma Control (GOAL) study was a large-scale, multinational randomized controlled trial that examined the feasibility of managing patients with uncontrolled asthma by aiming for total control (Bateman *et al.*, 2004). An economic appraisal of this clinical trial has recently been reported based on a statistical modeling approach to estimating cost-effectiveness (Briggs *et al.*, 2006). This section uses this previously reported economic analysis as a case study to demonstrate the potential advantages of this approach. A brief overview of the trial and the economic appraisal are given below. This is followed by an illustration and comparison of a 'conventional' approach to analyzing the trial in contrast to a statistical modeling approach. A final section offers some discussion.

9.3.1   The GOAL study

The GOAL study has recently been reported, and full details of the study design are published (Bateman *et al.*, 2004). Briefly, GOAL was a 52-week randomized controlled trial designed to assess the effectiveness of a predefined stepwise program of increased dosages of fluticasone propionate (FP) alone or in combination with the long-acting β_2-agonist, salmeterol, (SFC) in achieving asthma control. Control was defined using the definitions provided by the Global Initiative for Asthma/National Institutes of Health guidelines (Global Initiative for Asthma (GINA), 1998) as either 'totally controlled' or 'well controlled' based on a composite measure of asthma symptoms. Uncontrolled asthma was defined by the failure to meet either of the definitions of control. The unit of analysis for control was taken to be one week.

The study comprised two phases. In Phase I, following randomization, patients were in the dose-escalation phase where the dose of FP or SFC would be stepped up if they failed to achieve total control in at least seven weeks out of an eight-week assessment period. Patients entered the second maintenance phase, where their dose was kept constant if they had previously achieved total control in phase I or, if they remained uncontrolled, at the maximum dose in phase I. In total, 3416 patients were stratified into three approximately equal groups and were randomized between the FP and SFC arms of the trial. The three strata related to patients' use of inhaled corticosteroids prior to screening: stratum 1, no inhaled corticosteroid; stratum 2, 500 μg or less of beclomethasone dipropionate or equivalent; or stratum 3, more than 500–1000 μg or less of beclomethasone dipropionate or equivalent.

The economic analysis was conducted from the perspective of the UK National Health Service and the analysis was designed to meet the new UK National Institute for Health and Clinical Excellence (NICE) reference case (NICE, 2004). The general principle was to report an analysis based as closely as possible on the clinical trial. However, the trial did not collect information on quality of life utility values suitable for estimating cost per quality-adjusted life-years (QALYs) which are preferred by NICE for comparing the value in money of interventions across disease-areas. The trial did, however, collect information on a disease-specific quality of life instrument every four months. Therefore, the analysis makes use of external data providing an algorithm for linking the disease specific scale to utility values.

Estimating control. As described above, for each week of the trial, patients were classified in relation to the control of their asthma that they experienced. For the purposes of the economic analysis, patients were classified into four mutually exclusive control categories: 'totally controlled' (TC); 'well controlled' (WC); 'not well controlled' (NWC), but without exacerbation; and 'exacerbation' (X). The TC and WC categories were defined based on treatment guidelines. For those not achieving either of the control states in a given week, two categories were distinguished due to the important effect asthma exacerbations have on both resource use and health status—therefore weeks involving an exacerbation (X), defined as deterioration in asthma requiring treatment with an oral corticosteroid, or an emergency department visit, or hospitalisation, were distinguished from those where control was lacking, but no exacerbation was experienced (NWC).

Estimating costs. Three broad categories of health service resource use were collected as part of the clinical trial: secondary care contacts; primary care contacts and asthma medication costs. Secondary care information included: visits to emergency departments; length of time (no of days) in ICU; outpatient visits; and inpatient days. Primary care information included: general practitioner home visits during the day and the night; visits to the primary care clinic; and telephone calls to primary care clinic. Information on medications used were distinguished between study drugs (daily cost for each dosage level) and rescue medication use (per occasion cost). Unit costs relating to each of these resources (for the 2003–04 financial year) were taken from published sources for the UK (10). Once resource use values were weighted by their corresponding unit cost they were summed to provide a total cost on a weekly basis for each patient in the trial. Total costs per patient could then be obtained by simply adding over the 52-week trial follow-up.

Estimating QALYs. As mentioned above, the GOAL study did not collect utility data suitable for calculating QALYs. However, a disease specific measure, the Asthma Quality of Life Questionnaire (AQLQ)

Table 9.1 Regression model providing algorithm allowing calculation of EQ5D utilities from the AQLQ instrument (source: S. Macran and P. Kind, personal communication)

Explanatory variables	Coefficient	SE	p-value
Constant	0.19076	0.04853	<0.001
Activity limitations	0.09134	0.01281	<0.001
Symptoms	0.02089	0.01262	0.099
Age	−0.00013	0.00047	0.785
A13 Frustrated	0.00286	0.00818	0.727
A27 Afraid of getting out of breath	0.00166	0.00700	0.812

SE, standard error.

was collected. As part of a separate study, where data was collected on both the AQLQ and the EuroQol EQ5D instrument, an algorithm for obtaining an EQ5D utility score based on responses to some of the AQLQ domains was devised. The results of that algorithm (regression analysis) are presented in Table 9.1 (S. Macran and P. Kind, personal communication).

The regression model algorithm was employed to calculate a utility score for patients completing the AQLQ. The external algorithm provided a link between the AQLQ and the EQ5D utility tariff, and the regression of control status on EQ5D provided a link between EQ5D and the out-come of trial. In this way the QALYs associated with treatment were estimated allowing cost per QALY associated with treatment to be calculated.

9.3.2 Standard approach to estimating cost-effectiveness

In this section we outline what might be considered the standard approach to analysing the GOAL study, based on the approach that is commonly employed in economic analyses alongside trials (see Chapter 1) of calculating the mean costs and effects in each of the strata. The results from this approach are presented in Table 9.2 showing the costs and effectiveness (QALYs) in each arm together with the

Table 9.2 Cost-effectiveness analysis of GOAL performed separately for each stratum

	FP arm		SFC arm		SFC vs FP	
	Mean	SE	Mean	SE	Difference	95% limits
Stratum 1						
Total cost (£)	305	15	444	5	139	108, 170
Total QALYs	0.851	0.007	0.867	0.007	0.016	−0.003, 0.035
ICER (£/QALY)					8600	3800, dom
Stratum 1						
Total cost (£)	318	6	456	7	138	121, 155
Total QALYs	0.827	0.006	0.859	0.006	0.032	0.015, 0.049
ICER (£/QALY)					4300	2700, 9500
Stratum 1						
Total cost (£)	420	8	561	23	141	93, 190
Total QALYs	0.807	0.007	0.830	0.007	0.024	0.005, 0.042
ICER (£/QALY)					5900	2800, 28500

SE, standard error.
dom, SFC is dominated by FP.

incremental analysis separately for the three study strata. Confidence intervals for ICERs were estimated using Fieller's theorem, noting the negative (but low, $|\rho| < 0.1$) correlation between costs and effects.

On initial viewing the analysis in Table 9.2 appears to be perfectly adequate, giving as it does reasonable estimates of the quantities of interest, with only the upper limit of the ICER confidence interval on the first stratum being undefined (in the dominant quadrant of the CE plane). Nevertheless, by treating the strata totally independently, it is unclear to what extent any patterns are discernable from the data.

9.3.3 An alternative approach to estimating cost-effectiveness

In this section, an alternative approach to estimating cost effectiveness is taken based on modeling the components of the cost-effectiveness analysis. This is the approach that formed the basis of the published analysis and was developed in conjunction with the clinical steering

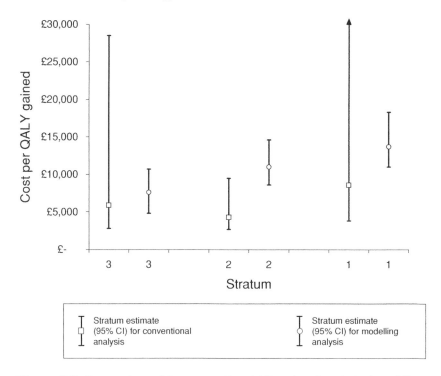

Figure 9.1 Comparison of the conventional (direct) and statistical modeling (indirect) approaches to estimating ICERs in GOAL by stratum

committee for the GOAL study, Briggs *et al* (2006). The overall model is made up of a linked set of four equations relating to two separate components of QALYs and two separate components of cost. Control status and treatment cost are modeled as a function of treatment arm and strata, and the HRQoL and other health service cost are modeled as a function of control status. These four separate equations are outlined in detail below, followed by the estimation of the cost-effectiveness.

Control status. To estimate the proportion of time patients spent in each category of control, while adjusting for the baseline strata and treatment allocation of each patient, the weekly observations of control status (TC, WC, NWC, X) was employed as the dependent variable in a multinomial regression model (Greene, 1993). The results of this model are presented in Table 9.3 showing the relative risks in

Table 9.3   Multinomial logistic regression for control status as a function of treatment group and strata (results shown as relative risks relative to the baseline category of NWC in stratum 1)

Explanatory variables	Relative risk	95% limits
TC		
Tx	1.951	1.711, 2.226
Stratum 2	0.649	0.555, 0.759
Stratum 3	0.394	0.334, 0.465
WC		
Tx	1.448	1.304, 1.608
Stratum 2	0.830	0.731, 0.944
Stratum 3	0.633	0.555, 0.722
X		
Tx	0.907	0.711, 1.157
Stratum 2	1.504	1.085, 2.084
Stratum 3	2.705	2.011, 3.638

the totally controlled, well controlled and exacerbation categories as a function of treatment allocation and stratum relative to the baseline category of an FP allocated patient in stratum 1 who is not well controlled. Robust standard errors are presented to account for the clustering of observations by individual subjects.

The results show that treatment with SFC increases the probability of being totally or well controlled relative to FP and reduces (although not significantly) the probability of having an exacerbation. Relative to stratum 1, patients in stratum 2 and 3 spend more time without control of asthma symptoms independent of treatment. The probability (or proportion of time) in each control state can be estimated by treatment allocation and by strata from the following expressions

$$\Pr(TC) = \frac{\exp\left\{X\beta^{TC}\right\}}{1 + \exp\left\{X\beta^{TC}\right\} + \exp\left\{X\beta^{WC}\right\} + \exp\left\{X\beta^{X}\right\}}$$

$$\Pr(WC) = \frac{\exp\left\{X\beta^{WC}\right\}}{1 + \exp\left\{X\beta^{TC}\right\} + \exp\left\{X\beta^{WC}\right\} + \exp\left\{X\beta^{X}\right\}}$$

$$\Pr(X) = \frac{\exp\left\{X\beta^X\right\}}{1 + \exp\left\{X\beta^{TC}\right\} + \exp\left\{X\beta^{WC}\right\} + \exp\left\{X\beta^X\right\}}$$

$$\Pr(NWC) = \frac{1}{1 + \exp\left\{X\beta^{TC}\right\} + \exp\left\{X\beta^{WC}\right\} + \exp\left\{X\beta^X\right\}}$$

where the superscript on the coefficients represents the different components of the multinomial regression equation as reported in Table 9.3. The predicted proportion of time over the 52 weeks of the trial that patients spent in the four control categories is shown in Table 9.4 by treatment arm and by baseline stratum. Treatment with SFC clearly has a positive impact on asthma control irrespective of the baseline stratum of the patient.

Resource use costs. Regression models were run separately for study treatment and other health service costs. For costs due to study treatment, allocation to the combination arm was expected to increase costs, but that the absolute difference would differ across the strata. Therefore, the regression model for treatment cost included treatment allocation and stratum as explanatory variables (including interaction effects between treatment and strata) and is reported in Table 9.5. The resulting predictions of (weekly) treatment cost are presented in Table 9.6, which show that SFC results in additional costs compared with FP alone.

Table 9.4   Predicted percentages of time spent in categories of control status by strata and treatment allocation

	FP arm				SFC arm			
	TC	WC	NWC	X	TC	WC	NWC	X
Stratum 1	31	32	37	1	42	32	26	0
Stratum 2	24	32	44	1	34	34	32	1
Stratum 3	17	29	52	2	26	32	40	1

FP, fluticasone propionate; SFC, salmeterol/fluticasone propionate in combination; TC, totally controlled; WC, well controlled; NWC, not well controlled (no exacerbation); X, exacerbation.

Table 9.5   Treatment cost regression by treatment allocation and stratum

	Coefficient	SE	p-value
Tx	3.31	0.130	<0.0001
Stratum2	0.62	0.128	<0.0001
Stratum3	2.19	0.128	<0.0001
TxS2	−0.54	0.181	0.003
TxS3	−1.26	0.181	<0.0001
Constant	4.98	0.092	<0.0001

SE, standard error.

The regression for other health service costs is reported in Table 9.7 and shows that control status was found to be a highly significant predicator of this type of cost ($p < 0.001$). Importantly, the addition of stratum and treatment allocation had no significant effect once control status was accounted for, suggesting that the use of SFC has no effect on other health service costs beyond the improvement in control status. Further, it was found that including a UK adjustment for the cost of exacerbation was highly significant ($p < 0.001$) suggesting that on average UK patients experiencing an exacerbation are treated less resource intensively (a reduced cost of £20). The predicted cost of a week in each control status category, with the cost of an exacerbation week adjusted to reflect the UK, are shown alongside the regression

Table 9.6   Predicted treatment cost by treatment allocation and by stratum

	FP arm		SFC arm	
	Cost (£)	SE	Cost (£)	SE
Stratum 1	4.98	0.09	8.29	0.09
Stratum 2	5.60	0.09	8.37	0.09
Stratum 3	7.17	0.09	9.21	0.09

SE, standard error; FP, fluticasone propionate; SFC, salmeterol/fluticasone propionate in combination.

Table 9.7 Other health service cost as a function of control status and adjusting for the UK experience

	Regression model		Weekly costs	
	Cost (£)	SE	Cost (£)	SE
TC (constant)	0.02	0.30	0.02	0.30
WC	0.14	0.25	0.16	0.28
NWC	1.09	0.28	1.11	0.28
X	52.35	0.89	32.29	5.62
X * UK	−20.08	5.68		

SE, standard error; TC, totally controlled; WC, well controlled; NWC, not well controlled (no exacerbation); X exacerbation; UK, United Kingdom indicator variable.

results in Table 9.7. It is clear from these results that cost decreases with improved control and the most significant health care costs are associated with exacerbations.

Health-related quality of life and QALYs. The approach to estimating utilities for the calculation of QALYs mirrors the regression analysis of other cost above. The utility values mapped from the AQLQ scores formed the explanatory variable in a regression with control status as explanatory. The results of this regression model are presented in Table 9.8 which includes a dummy variable for UK patients (all explanatory variables were significant predictors of quality of life at $p < 0.001$). In

Table 9.8 Health-related quality of life utilities by control status

	Regression model		Weekly utilities	
	Utility	SE	Utility	SE
TC (constant)	0.902	(0.003)	0.946	(0.011)
WC	−0.045	(0.002)	0.900	(0.011)
NWC	−0.104	(0.003)	0.842	(0.011)
X	−0.216	(0.007)	0.729	(0.013)
UK	0.044	(0.011)		

TC, totally controlled; WC, well controlled; NWC, not well controlled (no exacerbation); X, exacerbation; UK, United Kingdom indicator variable.

contrast to the other health service cost regression, adding in treatment as an explanatory variable resulted in a marginally significant utility gain of 0.01 ($p = 0.044$) even when controlling for control status, suggesting there may be additional benefits of treatment not captured by the simple categorization of control used in the trial. In the analysis presented here, this additional treatment benefit is ignored, making the analysis conservative with respect to the value of the combination product. Similarly, stratum 3 was associated with a small, but significant ($p < 0.001$), reduction in utility of -0.02 which is not accounted for in the analysis presented here.

Estimating cost-effectiveness and associated uncertainty. To estimate cost-effectiveness for each of the strata in the GOAL study the predictions from the four models presented in Tables 9.4, 9.6, 9.7 and 9.8 were combined. The mechanics of this approach are illustrated in Table 9.9 which presents all of the information from the previous tables by control status, stratum and allocation arm. Costs and QALYs are estimated within each stratum and by allocation arm through a process of weighted averaging over control status. The final column of Table 9.9 summarizes the differences between the arms.

Although the standard errors of the regression coefficients reflect the uncertainty in the estimation of coefficients, such that the calculation of confidence limits for predictions from individual regressions is straightforward, the calculation of confidence limits for the derived incremental costs and effects in Table 9.9 is less straightforward. While the covariance matrices of individual regressions could be employed to correctly characterize correlation between parameters estimates of an individual equation, such an approach would not capture potential correlation between parameters estimates in the different equations. To capture this potential correlation, the non-parametric approach of bootstrapping was employed. This involved bootstrapping a resample of the entire dataset (at the patient level), fitting all of the equations on the resampled data set and calculating the corresponding cost-effectiveness results. This process was repeated 1000 times in total to allow accurate estimation of the confidence intervals to represent uncertainty. The incremental analysis and cost-per-QALY calculations are reproduced in Table 9.10 using the information from Table 9.9, but also presenting confidence intervals for each reported

Table 9.9 Estimating the incremental cost and QALYs by stratum

| | FP arm | | | | | SFC qrm | | | | | |
	TC	WC	NWC	X	Weighted average	TC	WC	NWC	X	Weighted average	Difference
Stratum 1											
Status distribution (%)	31	32	37	1		42	32	26	0		
Treatment cost (£)	4.98	4.98	4.98	4.98	4.98	8.29	8.29	8.29	8.29	8.29	3.31
Other health care costs (£)	0.02	0.16	1.11	32.29	0.63	0.02	0.16	1.11	32.29	0.45	−0.18
HRQoL/QALYs	0.946	0.900	0.842	0.729	0.892	0.946	0.900	0.842	0.729	0.903	0.012
Stratum 2											
Status distribution (%)	24	32	44	1		34	34	32	1		
Treatment cost (£)	5.60	5.60	5.60	5.60	5.60	8.37	8.37	8.37	8.37	8.37	2.77
Other health care costs (£)	0.02	0.16	1.11	32.29	0.83	0.02	0.16	1.11	32.29	0.61	−0.22
HRQoL/QALYs	0.946	0.900	0.842	0.729	0.884	0.946	0.900	0.842	0.729	0.896	0.012
Stratum 3											
Status distribution (%)	17	29	52	2		26	32	40	1		
Treatment cost (£)	7.17	7.17	7.17	7.17	7.17	9.21	9.21	9.21	9.21	9.21	2.05
Other health care costs (£)	0.02	0.16	1.11	32.29	1.25	0.02	0.16	1.11	32.29	0.95	−0.31
HRQoL/QALYs	0.946	0.900	0.842	0.729	0.874	0.95	0.90	0.84	0.73	0.886	0.012

FP, fluticasone propionate; SFC, salmeterol/fluticasone propionate in combination; TC, totally controlled; WC, well controlled; NWC, not well controlled (no exacerbation); X, exacerbation.

value derived from the bootstrap analysis. To facilitate comparison with the direct estimates of cost-effectiveness from Section 9.3, the ICERs by stratum with associated confidence intervals are presented in Table 9.1.

9.3.4   Comparing the two analyses of GOAL

The two approaches to analyzing the same data from the GOAL study are starkly contrasted in Table 9.10. The conventional approach based on splitting the data leads to much wider confidence intervals than does the approach based on a series of statistical equations for the components of the economic analysis that adjust for treatment allocation and stratum. By modeling the effect of the stratum directly, the full power of the data is retained. Nevertheless, this increased power comes through the extra structural assumptions imposed on the analysis and the credibility of the results is only as good as the

Table 9.10   Cost-per-QALY by stratum of GOAL estimated using the indirect approach

		Difference	
	Point estimate	lower 95% limit	Upper 95% limit
Stratum 1			
Total costs (£)	163	147	177
Total QALYs	0.0118	0.0094	0.0143
ICER (£)	13 700	11 000	18 300
Stratum 2			
Total costs (£)	132	114	147
Total QALYs	0.0120	0.0094	0.0145
ICER (£)	11 000	8600	14 600
Stratum 3			
Total costs (£)	90	61	109
Total QALYs	0.0118	0.0093	0.0141
ICER (£)	7600	4800	10 700

assumptions underlying the modeling. For example, the results from the cost regression analyses suggest the important predictors of costs differ between treatment costs and other health service costs. For treatment costs, it is the trial allocation arm and baseline stratum that are the important predictors of cost. By contrast, these variables are not predictive of other health service costs once control status is added to the regression. The results for the health-related quality of life regression are less definitive. In particular, it was reported in the text that both treatment allocation and stratum 3 were significantly predictive of health-related quality of life independently of control status. Nevertheless, in the reported cost-effectiveness analysis, these modest additional effects of treatment and stratum 3 were excluded in order to simplify the modeling assumptions.

An additional issue is that the GOAL study was a multinational trial, but that the focus of this analysis was the cost-effectiveness in the UK. The approach used was to employ the whole data set on resource use in order to maximize the power of the resulting analysis, but to employ a UK indicator variable to adjust the analysis for UK-specific effects. This approach highlighted that, on average, the cost of treating an exacerbation in the UK was £20 below the overall cost in the trial.

The use of the non-parametric bootstrap allows confidence intervals to be calculated for the estimates of cost-effectiveness based on combining a series of statistical relationships estimated from the data. These confidence intervals represent uncertainty related to sampling variation and include potential correlation structure between the estimated equations. However, the confidence intervals do not include uncertainty related to the algorithm that links the AQLQ disease specific instrument to the utility scores suitable for calculating QALYs, since this algorithm was based on data external to the trial.

9.4   SUMMARY

In this chapter an alternative approach to obtaining incremental costs and effects for the calculation of cost-effectiveness was outlined based on a series of statistical models of the components of costs and effects. Through the use of a case study of an economic evaluation of a large, stratified clinical trial of asthma treatment, the potential advantages of such an indirect approach to estimating increment costs and effects were outlined in comparison to the conventional approach based on the direct estimation of mean costs and effects outlined earlier in this book. The principle advantage of the modeling approach is the ability to improve the potential power of a study by avoiding the need to 'split' data for different sub-groups and to directly model the components of cost and effectiveness more efficiently. In addition, such an approach can help to handle missing data and avoid some of the issues associate with rare, but high-cost, events. Furthermore, such an approach may help readers and analysts to understand the mechanism and nature of the effects of treatment and other important prognostic factors. Nevertheless, the credibility of cost-effectiveness results generated in this way rests on the credibility of the statistical modeling assumptions. Some commentators are likely to argue that only the direct approach provides for the full unbiased analysis associated with randomized experiments.

The case study employed in this chapter was slightly unusual in that cost-effectiveness was assessed over the trial period. This is due to the fact that, in this patient group, treatment is considered to only have an effect on HRQoL not length of life. More commonly, treatment may be expected to affect mortality and therefore a full economic analysis should be based on a lifetime analysis. The statistical modeling framework proposed in this chapter can also be used in such a situation to parameterize survival data with the aim of projecting survival curves forward beyond the period of follow-up. This approach was used in the development of a disease model for diabetes based on the UK Prospective Diabetes Study, the longest ever trial of diabetes treatment, which nevertheless required statistical modeling of the full course of the disease history to facilitate economic modeling (Clarke *et al.*, 2004).

In practice, the advantages of statistically modeling data sources must be weighed against the additional structural assumptions they bring to the analysis. As with all modeling studies, the key is to ensure that assumptions underlying the model are explicit to allow the reader to judge their validity.

References

Altman DG. (1985) Comparability of randomisd groups. *Statistician*, **34**, 125–136.

The ASSENT-3 Investigators. (2001) Efficacy and safety of tenecteplase in combination with enoxaparin, abciximab, or unfractionated heparin: the ASSENT-3 randomised trial in acute myocardial infarction. *Lancet*, **358**, 605–613.

Bang H and Tsiatis AA. (2000) Estimating medical costs with censored data, *Biometrika*, **87**, 329–343.

Bateman ED, Boushey HA, Bousquet J, Busse WW, Clark TJH, Pauwels RA, Pedersen SE and the GOAL Investigators Group (2004) Can guideline-defined asthma control be achieved?: The gaining optimal asthma control study", *American Journal of Respiratory and Critical Care Medicine*, **170**, 836–844.

Black WC. (1990) The CE plane: a graphic representation of cost-effectiveness, *Medical Decision Making*, **10**, 212–214.

Bloomfield DJ, Krahn MD, Neogi T, Panzarella T, Warde P, Willan AR, Ernst S, Moore MJ, Neville A, Tannock IF. (1998) Economic evaluation of chemotherapy with mitoxantrone plus prednisone for symptomatic hormone-resistant prostate cancer: Based on a Canadian trial with palliative endpoints, *Journal of Clinical Oncology*, **16**, 2272–2279.

Briggs AH. (1999) A Bayesian approach to stochastic cost-effectiveness analysis, *Health Economics*, **8**, 257–261.

Briggs A, Bousquet J, Wallace M, Busse WW, Clark TJH, Pederson SE, Bateman E. (2006) Cost effectiveness of asthma control: an economic appraisal of the GOAL study. *Allergy*, **61**, 531–536.

Briggs AH and Fenn P. (1998) Confidence intervals or surfaces? Uncertainty on the cost-effectiveness plane, *Health Economics*, **7**, 723–740.

Briggs AH and Gray AM. (1998a) The Distribution of Health Care Costs and Their Statistical Analysis for Economic Evaluation, *Journal of Health Services Research and Policy*, **3**, 233–245.

Briggs AH and Gray AM. (1998b) Power and sample size calculations for stochastic cost-effectiveness analysis. *Medical Decision Making*, **18**, S81–S92.

Briggs AH and Gray AM. (1999) Handling uncertainty when performing economic evaluation of health care interventions, *Health Technology Assessment*, vol. 3, no. 2.

Briggs AH, Mooney CZ and Wonderling DE. (1999) Constructing confidence intervals for cost-effectiveness ratios: an evaluation of parametric and non-parametric techniques using monte carlo simulation, *Statistics in Medicine*, **18**, 3245–3262.

Briggs A, Nixon R, Dixon S and Thompson S. (2005) Parametric modelling of cost data: some simulation evidence, *Health Economics*, **14**, 421–428.

Briggs AH, Sculpher MJ and Buxton MJ. (1994) Uncertainty in the economic evaluation of health care technologies: the role of sensitivity analysis, *Health Economics*, **3**, 95–104.

Briggs AH, Wonderling DE and Mooney CZ. (1997) Pulling cost-effectiveness analysis up by its bootstraps: a non-parametric approach to confidence interval estimation, *Health Economics*, **6**, 327–340.

Brown HK and Kempton RA. (1994) The application of REML in clinical trials. *Statistics in Medicine*, **13**, 1601–1617.

Chaudhary MA and Stearns SC. (1996) Confidence intervals for cost-effectiveness ratios: An example from a randomized trial, *Statistics in Medicine*, **15**, 1447–1458.

Chiba N, van Zanten SJ, Sinclair P, Ferguson RA, Escobedo S, Grace E. (2002) Treating Helicobacter pylori infection in primary care patients with uninvestigated dyspepsia: the Canadian adult dyspepsia empiric treatment-Helicobacter pylori positive (CADET-Hp) randomised controlled trial. *BMJ*, **324**, 1012–1016.

Clarke PM, Gray AM, Briggs, A, Farmer AJ, Fenn, P, Stevens RJ, Matthews DR, Stratton IM, and Holman, R. (2004) A model to estimate the lifetime health outcomes of patients with type 2 diabetes: the United Kingdom Prospective Diabetes Study (UKPDS) Outcomes Model (UKPDS no. 68), *Diabetologia*, **47**, 1747–1759.

Claxton K. The irrelevance of inference: a decision-making approach to the stochastic evaluation of health care technologies. (1999) *Journal of Health Economics*, **18**, 341–364.

Claxton K and Posnett J. (1996) An economic approach to clinical trial design and research priority setting. *Health Economics*, **5**, 513–524.

Cochran WG. (1977) *Sampling techniques*, John Wiley & Sons, New York.

Collins R, Gray R, Godwin J, Peto R. (1987) Avoidance of large biases and large random errors in the assessment of moderate treatment effects: the need for systematic overviews. *Statistics in Medicine*, **6**, 245–254.

Commonwealth of Australia. (1990) *Guidelines for the pharmaceutical industry on preparation of submissions to the Pharmaceutical Benefits Advisory Committee: including submissions involving economic analyses.* Woden (ACT) Department of Health, Housing and Community Services.

Connolly SJ, Gent M, Roberts RS, Dorian P, Roy D, Sheldon RS, Mitchell LB, Green MS, Klein GJ and O'Brien B. (2000) Canadian implantable defibrillator study (CIDS): a randomized trial of the implantable cardioverter defibrillator against amiodarone. *Circulation*, **101**, 1297–1302.

Cook JR, Drummond M, Glick H and Heyse JF. (2003) Assessing the appropriateness of combining economic data from multinational clinical trials. *Statistics in Medicine*, **22**, 1955–1976.

Dahlof B, Lindholm LH, Hansso, L, Scherste, B, Ekbom T, and Wester PO. (1991), Morbidity and mortality in the Swedish Trial in Old Patients with Hypertension (STOP-Hypertension), *Lancet*, **338**, 1281–1285.

Dersimonian R and Laird N. (1986) Meta-analysis in clinical trials. *Controlled Clinical Trials*, **7**, 177–188.

Detsky AS. (1993) A guideline for preparation of economic analysis of pharmaceutical products: a draft document for Ontario and Canada. *Pharmacoeconomics* **3**, 354–361.

Drummond MF, O'Brien BJ, Stoddart GL and Torrance GW. (1997) *Methods for the Economic Evaluation of Health Care Programmes*, 2nd ed. Oxford: Oxford University Press.

Duan N. (1983) Smearing Estimate: A Nonparametric Retransformation Method. *Journal of the American Statistical Association*, **78**, 605–610.

Efron B. (1987) Better bootstrap confidence intervals, *Journal of the American Statistical Association*, **82**, 171–200.

Efron B and Morris C. (1972) Empirical Bayes on vector observations: An extension of Stein's method. *Biometrika*, **59**, 335–347.

Efron B and Tibshirani R. (1993) *An Introduction to the Bootstrap*, Chapman and Hall, New York.

Fenwick E, O'Brien BJ, Briggs AH. (2004) Cost-effectiveness acceptability curves – facts, fallacies and frequently asked questions, *Health Economics*, **13**, 405–415.

186   *References*

Gail M and Simon R. (1985) Testing for Qualitative Interactions between Treatment Effects and Patient Subsets, *Biometrics*, **41**, 361–372

Gardiner JC, Huebner M, Jetton J and Bradley CJ. (2000) Power and sample size assessments for tests of hypotheses on cost-effectiveness ratios. *Health Economics*, **9**, 227–234.

Garry R, Fountain J, Mason S, et al. (2004) The EVALUATE study: two parallel randomised trials, one comparing laparoscopic with abdominal hysterectomy, the other comparing laparoscopic with vaginal hysterectomy. *BMJ* 2004, **328**, 129–133.

Glasziou PP and Irwig LM. (1995) An evidence based approach to individualising treatment, *British Medical Journal*, **311**, 7016, 1356–1359.

Global Initiative for Asthma (GINA) (1998) *Pocket guide for asthma management and prevention*, Natiaonal Institutes for Health, National Heart, Lung and Blood Institute, Bethesda, Publication No. 95-3659-B.

Greene WH. (1993) Econometric Analysis (2nd Ed). MacMillan: New York; 486–489.

Grieve R, Nixon RM, Thompson SG and Normand C. (2005) Using multilevel models for assessing the variability of multinational resource use and cost data. *Health Economics*, **14**, 185–196.

Grundy PM, Healy MJR and Rees DH. (1956) Economic choice of the amount of experimentation. *Journal of the Royal Statistical Society, Series A*, **18**, 32–48.

Hanley N, Ryan M, and Wright R. (2003) Estimating the monetary value of health care: lessons from environmental economics, *Health Economics*, **12**, 3–16.

Hannah ME, Hannah WJ, Hewson SH, Hodnett ED, Saigal S and Willan AR. (2000) Term Breech Trial: a multicentre international randomised controlled trial of planned caesarean section and planned vaginal birth for breech presentation at term. *The Lance*, **356**, 1375–1383.

Hanemann WM. (1991) Willingness to pay and willingness to accept: How much can they differ? *The American Economic Review*, **18**, 635–647.

Hardy RJ and Thompson SG. (1998) Detecting and describing heterogeneity in meta-analysis. *Statistics in Medicine*, **17**, 841–885.

Heitjan DF (2000), Fieller's method and net health benefits. *Health Economics*, 9, 327–335.

Hoch JS, Briggs AH, Willan AR. (2002) Something old, something new, something borrowed, something blue: a framework for the marriage of health econometrics and cost-effectiveness analysis. *Health Economics*, **11**, 415–430.

Hutton EK, Kaufman K, Hodnett E, Amankwah K, Hewson SA, McKay D, Szalai JP and Hannah ME. (2003) External cephalic version beginning at 34 weeks' gestation versus 37 weeks' gestation: a randomized multicenter trial. *American Journal of Obstetrics and Gynecology*, **189**, 245–254.

Johannesson M, Dahlof B, Lindholm LH, Ekbom T, Hansson L, Oden A, Schersten B, Wester PO, Jonsson B. The cost-effectiveness of treating hypertension in elderly people–an analysis of the Swedish Trial in Old Patients with Hypertension (STOP Hypertension), *Journal of Internal Medicine*, **234**, 317–323.

Johansson PO. (1995) Evaluating Health Risks: An Economic Perspective. Cambridge University Press: Cambridge; 38.

Johnson FR, Desvousges WJ, Ruby MC, Stieb D and De Civita P. (1998) Eliciting stated health preferences: an application to willingness to pay for longevity, *Medical Decision Making*, **18 Suppl**, S57–S67.

Jonsson B, Johannesson M, Kjekshus J, Olsson AG, Pedersen TR, and Wedel H. (1996) Cost-effectiveness of cholesterol lowering. Results from the Scandinavian Simvastatin Survival Study (4S), *European Heart Journal*, **17**, 1001–1007.

Kahneman D and Tversky A. (1979) Prospect theory: An analysis of decision under risk. *Econometrica*, **47**, 263–291.

Kind P. (1996) The EuroQol instrument: an index of health-related quality of life. In *Quality of Life and Pharmacoeconomics in Clinical Trials, 2nd edition*, Spilker B (ed). Lippincott-Raven: Philadelphia, 191–201.

Koopmanschap MA, Touw KCR and Rutten FFH. (2001) Analysis of cost and cost-effectiveness in multinational trials. *Health Policy*, **58**, 175–186.

Kuntz KM and Weinstein MC. (2001) Modeling in economic evaluation. In *Economic Evaluation in Health Care: Merging Theory with Practice*, MF Drummond & A McGuire, (eds), Oxford University Press: Oxford, 141–171.

Lachin JM. (1981) Introduction to sample size determination and power analysis for clinical trials, *Controlled Clinical Trials*, **2**, 93–113.

Laska EM, Meisner M and Siegel C. (1997) Statistical inference for cost-effectiveness ratios, *Health Economics*, **6**, 229–242.

Laupacis A, Feeny D, Detsky A and Tugwell P. (1992) How attractive does a new technology have to be to warrant adoption and utilization? Tentative guidelines for using clinical and economic evaluations, *Canadian Medical Association Journal*, **146**, 472–481.

Leaf A. (1989) Cost-effectiveness as a criterion for Medicare coverage, *New England Journal of Medicine*, **321**, 898–900.

Lehman AF, Dixon LB, Kernan E *et al.* (1997) A randomized trial of assertive community treatment for homeless persons with severe mental illness, *Archives of General Psychiatry*, **54**, 1038–1043.

Lehman AF, Dixon LB, Hoch JS *et al.* (1999) Cost-effectiveness of assertive community treatment for homeless persons with severe mental illness, *British Journal of Psychiatry*, **174**, 346–352.

Lemeshow S, Hosmer Jr. DW, Klar J and Lwanga SK. (1990) The adequacy of sample size in health studies. John Wiley and Sons, New York.

Lin DY, Feuer EJ, Etzioni R and Wax Y. (1997) Estimating medical costs from incomplete follow-up data, *Biometrics*, **53**, 419–434.

Lin DY. (2000) Linear regression analysis of censored medical costs, *Biostatistics*, **1**, 35–47.

Lumley T, Diehr P, Emerson S and Chen L. (2002) The importance of the normality assumption in large public health data sets, *Annual Review of Public Health*, **23**, 151–169.

Manca A, Hawkins N, Sculpher MJ. (2005a) Estimating mean QALYs in trial-based cost-effectiveness analysis: the importance of controlling for baseline utility. *Health Economics*, **14**, 487–496.

Manca A, Rice N, Sculpher MJ and Briggs AH. (2005b). Assessing generalisability by location in trial-based cost-effectiveness analysis: the use of multilevel models. *Health Economics*, **14**, 471–485.

Manning WG and Mullahy J. (2001) Estimating log models: to transform or not to transform? *Journal of Health Economics*, **20**, 461–494.

McAlister FA. (2002) Commentary: relative treatment effects are consistent across the spectrum of underlying risks . . . usually, *International Journal of Epidemiology*, **31**, 76–77.

Mooney CZ and Duval RD. (1993) *Bootstrapping: A Nonparametric Approach to Statistical Inference*. Sage, Newbury Park, CA.

Morrison GC. (1998) Understanding the disparity between WTP and WTA: endowment effect, substitutability, or imprecise preferences? *Economics Letters* **59**, 189.

National Institute for Clinical Excellence (NICE) 2004, *Guide to the methods of technology assessment* NICE, London.

Nixon RM and Thompson SG. (2005) Incorporating covariate adjustment, subgroup analysis and between-centre differences into cost-effectiveness evaluations. *Health Economics*, **14**, 1217–1229.

O'Brien BJ. (1996) Economic evaluation of pharmaceuticals: Frankenstein's monster or vampire of trials? *Medical Care*, **34**, DS99–DS108.

O'Brien BJ. (1997) A tale of two (or more) cities: geographic transferability of pharmacoeconomic data. *American Journal of Managed Care*, **3**, S33–S40.

O'Brien BJ, Connolly SJ, Goeree R, Blackhouse G, Willan AR, Yee R, Roberts RS and Gent M. (2000) Cost-effectiveness of the implantable cardioverter defibrillator results from the Canadian implantable defibrillator study (CIDS). *Circulation*, **103**, 1416–1421.

O'Brien BJ, Drummond MF, Labelle RJ and Willan AR. (1994) In search of power and significance: Issues in the design and analysis of stochastic cost-effectiveness studies in health care, *Medical Care*, **32**, 150–163.

O'Brien BJ and Gafni A. (1996) When do the dollars make sense? Toward a conceptual framework for contingent valuation studies in health care, *Medical Decision Making*, **16**, 288–299.

O'Brien BJ, Gertsen K, Willan AR and Faulkner L. (2002) Is there a kink in consumers' threshold value for cost-effectiveness in health care? *Health Economics*, **11**, 175–180.

O'Hagan A and Stevens JW. (2001) Bayesian assessment of sample size for clinical trials of cost-effectiveness. *Medical Decision Making*, **21**, 219–230.

O'Hagan A and Stevens JW. (2003) Assessing and comparing costs: how robust are the bootstrap and methods based on asymptotic normality? *Health Economics*, **12**, 33–49.

Oxman AD, Guyatt GH. (1992) A consumer's guide to subgroup analyses. *Annals of Internal Medicine*, **116**, 78–84.

Pauly MV. (1995) Valuing health care benefits in money terms. In Sloan FA (ed). *Valuing Health Care*, Cambridge University Press, Cambridge.

Pezeshk H. (2003) Bayesian techniques for sample size determination in clinical trials: a short review. *Statistical Methodology in Medical Research* 2003, **12**, 489–504.

Pinto EM, Willan AR and O'Brien BJ. (2005) Cost-effectiveness analysis for multinational clinical trials. *Statistics in Medicine*, **24**, 1965–1982.

Pocock SJ. (1984) *Clinical Trials: A Practical Approach*. John Wiley & Sons Ltd., Chichester.

Raiffa H and Schlaifer R. (2000) *Applied Statistical Decision Theory*. John Wiley and Sons, New York.

Raikou M, Gray A, Briggs A, Steven, R, Cull C, McGuire A, Fenn P, Stratton I, Holman, R, and Turner R. (1998) Cost effectiveness analysis of improved blood pressure control in hypertensive patients with type 2 diabetic patients (HDS7): UKPDS 40, *British Medical Journal*, **317**, 720–726.

Reed SD, Anstrom KJ, Bakhai A, Briggs AH, Califf RM, Cohen DJ, Drummond MF, Glick HA, Gnanasakthy A, Hlatky MA, O'Brien BJ, Torti FM, Tsiatis AA, Willan AR, Mark DB and Schulman KA. (2005) Conducting economic

190   *References*

evaluations alongside multinational clinical trials: Toward a research consensus. *American Heart Journal*, **149**, 434–443.

Scandinavian Simvastatin Survival Study Group (1994) Randomised trial of cholesterol lowering in 4444 patients with coronary heart disease: the Scandinavian Simvastatin Survival Study (4S). *Lancet*, **344**, 1383–1389.

Sculpher M, Manca A, Abbott J, Fountain J, Mason S and Garry R. (2003) The cost-effectiveness of laparoscopic-assisted hysterectomy in comparison of standard hysterectomy: the EVALUATE trial. *BMJ*, 2004, **328**, 134–137.

Senn S. (1997) *Statistical Issues in Drug Development*. John Wiley & Sons, Chichester.

Skene AM and Wakefield JC. (1990) Hierarchical models for multicentre binary response studies. *Statistics in Medicine*, **9**, 919–929.

Spiegelhalter, DJ, Thomas, A and Best, NG (1999). *WinBUGS Version 1.2 User Manual*. Cambridge: MRC Biostatistics Unit.

Sutton AJ, Cooper NJ, Abrams KR, Lambert PC and Jones DR. (2005) A Bayesian approach to evaluating net clinical benefit allowed for parameter uncertainty. *Journal of Clinical Epidemiology*, 58, 26–40.

Tannock IF, Osoba D, Stockler MR, Ernst DS, Neville AJ, Moore MJ, Armitage GR, Wilson JJ, Venner PM, Coppin CM, Murphy KC. (1996) Chemotherapy with mitoxantrone plus prednisone or prednisone alone for symptomatic hormone-resistant prostate cancer: A Canadian trial with palliative endpoints. *Journal of Clinical Oncology*, **14**, 1756–1764.

Thompson SG and Barber JA. (2000) How should cost data in pragmatic randomised trials be analysed? *BMJ*, **320**, 1197–1200.

Torrance GW. (1976) Health status index models: a unified mathematical view, *Management Science*, **22**, 990–1001.

Torrance GW. (1986) Measurement of health state utilities for economic appraisals: a review, *Journal of Health Economics*, **5**, 1–30.

Torrance GW. (1987) Utility approach to measuring health related quality of life, *Journal of Chronic Disease*, **40**, 593–600.

van Hout BA, Al MJ, Gordon GS, and Rutten FF. (1994) Costs, effects and C/E ratios alongside a clinical trial, *Health Economics*, **3**, 309–319.

Weinstein MC and Stason WB. (1977) Foundation of cost-effective analysis for health and medical practices, *New England Journal of Medicine*, **296**, 716–721.

Willke RJ, Glick HA, Polsky D and Schulman K. (1998) Estimating country-specific cost-effectiveness from multinational clinical trials. *Health Economics*, **7**, 481–493.

Willan AR. (1994) Power function arguments in support of an alternative approach for analyzing management trials. *Controlled Clinical Trials*, **15**, 211–219.

Willan AR. (2004) Incremental net benefit in the analysis of economic data from clinical trials with application to the CADET-Hp Trial. *European Journal of Gastroenterology and Hepatology*, **16**, 543–549.

Willan AR, Briggs AH and Hoch JS. (2004) Regression methods for covariate adjustment and subgroup analysis for non-censored cost-effectiveness data, *Health Economics*, **13**, 461–475.

Willan AR, Chen EB, Cook RJ and Lin DY. (2003) Incremental net benefit in randomized clinical trials with qualify-adjusted survival. *Statistics in Medicine*, **22**, 353–362.

Willan AR, Cruess AF and Ballantyne M. (1996) Argon green vs krypton red laser photocoagulation of extrafoveal choroidal neovascular lesions: Three-year results in age-related macular generation. *Canadian Journal Ophthalmology*, **31**, 11–17.

Willan AR and Lin DY. (2001) Incremental net benefit in randomized clinical trials. *Statistics in Medicine*, **20**, 1563–1574.

Willan AR, Lin DY, Cook RJ and Chen EB. (2002) Using inverse-weighing in cost-effectiveness analysis with censored data, *Statistical Methods in Medical Research*, **11**, 539–551.

Willan AR, Lin DY and Manca A. (2005) Regression methods for cost-effectiveness analysis with censored data. *Statistics in Medicine*, **24**, 131–145.

Willan AR and O'Brien BJ. (1996) Confidence intervals for cost-effectiveness ratios: An application of Fieller's theorem. *Health Economics*, **5**, 297–305.

Willan AR and O'Brien BJ. (1999) Sample size and power issues in estimating incremental cost-effectiveness ratios form clinical trials data. *Health Economics*, **8**, 203–211.

Willan AR and O'Brien BJ. (2001) Cost prediction models for the comparison of two groups. *Health Economics*, **10**, 363–366.

Willan AR, O'Brien BJ and Leyva RA. (2001) Cost-effectiveness analysis when the WTA is greater than the WTP. *Statistics in Medicine*, **20**, 3251–3259.

Willan AR, Pinto EM. (2005) The value of information and optimal clinical trial design, *Statistics in Medicine*, **24**, 1791–1806.

Willan AR, Pinto EM, O'Brien BJ, Kaul P, Goeree R, Lynd L and Armstrong PW. (2005) Country-specific cost comparisons from multinational clinical trials using empirical Bayesian shrinkage estimation: The Canadian ASSENT-3 Economic Analysis. *Health Economics*, **14**, 327–338.

Zethraeus N and Lothgren M. (2000) *On the equivalence of the net benefit and the Fieller's methods for statistical inference in cost-effectiveness analysis.* Working paper no. 379, SSE/EFI Working Series Papers in Economics and Finance, Stockholm School of Economics.

Zhao H and Tian L. (2001) On estimating medical cost and incremental cost-effectiveness ratios with censored data, *Biometrics*, **57**, 1002–1008.

Author Index

Altman DG 15, 25, 117, 143

Bang H 31, 34
Bateman ED 167, 168
Best NG 164
Black WC 5
Bloomfield 72
Briggs AH 1, 11, 13, 14, 27, 43, 45, 46,
    48, 52, 54, 56, 67, 88, 93, 94, 95,
    96, 117, 119, 142, 145, 165, 167
Brown HK 146, 162

Chaudhary MA 45, 46, 54
Chiba PM 68
Clarke PM 181
Claxton K 104, 109, 110
Cochran WG 13, 48
Collins R 143
Connolly SJ 77
Cook JR 146, 152, 153

Dahlof B 88
Dersimonian R 155
Detsky AS 1
Drummond MF 5
Duan N 16

Efron B 55, 56, 156

Fenn P 46, 52, 54, 96
Fenwick E 54

Gafni A 7
Gail M 152
Gardiner JC 94, 101
Garry R 82
Glasziou PP 166
Greene WH 177, 119, 172
Grieve R 146, 162
Gray AM 13, 88, 94, 95, 96
Grundy PM 109
Guyatt GH 143

Hanemann WM 60
Hanley N 7
Hannah ME 108
Hardy RJ 157
Heitjan DF 51
Hoch JS 14, 117, 119, 142
Hutton EK 114

Johannesson M 88
Johansson PO 60
Johnson FR 7
Jonsson B 152

Kahneman D 60
Kind P 80, 172
Koopmanschap MA 146
Kuntz KM 165

Lachin JM 104
Laska M 46

Statistical Analysis of Cost-effectiveness Data. A. Willan and A. Briggs
© 2006 John Wiley & Sons, Ltd.

Laupacis A 100
Leaf A 1
Lehman AF 121
Lemeshow S 104
Lin DY 9, 28, 29, 31, 34, 42, 118, 130, 132, 134
Lothgren M 51
Lumley T 14, 48

Manca A 25, 118, 130, 134, 143, 146, 162
Manning WG 13
McAlister FA 148
Mooney CZ 55, 56
Morrison GC 60
Mullahy J 13

Nixon RM 13, 146, 162, 163

O'Brien BJ 1, 7, 14, 20, 45, 46, 48, 60, 62, 77, 94, 95, 96, 98, 103, 146, 147, 156
O'Hagan A 13, 106, 107
Oxman AD 143

Pauly MV 7
Pezeshk H 109
Pinto EM 109, 110, 114, 146, 156, 157, 160
Pocock SJ 115, 117, 143

Raiffa H 109

Raikou M 86
Reed D 149, 150, 164

Schlaifer R 109
Sculpher M 82
Simon R 152
Skene AM 146, 162
Spiegelhalter DJ 164
Stevens JW 13, 106, 107
Sutton AJ 166

Tannock IF 72
Thomas A 164
Thompson SG 13, 48, 146, 157, 162, 163
Tian L 31, 34
Tibshirani R 55, 56
Torrance GW 4, 22
Tsiatis AA 31, 34

van Hout BA 52

Weinstein MC 22, 165, 166
Willke RJ 146, 151
Willan AR 1, 11, 14, 20, 27, 31, 34, 35, 37, 38, 40, 41, 43, 46, 48, 62, 67, 68, 93, 94, 98, 103, 104, 108, 109, 110, 114, 117, 118, 119, 130, 134, 142, 145, 146, 156, 165

Zethraeus N 51
Zhao H 31, 34

Subject Index

absolute risk reduction 3
aggregate level analysis 146, 154, 156, 157
area under the curve 23, 36, 39

Bayesian methods 9, 43, 52, 57, 67, 76, 86, 93, 106, 110
Bayesian sample size 106
bias 2, 16, 25, 27, 39, 56
bias corrected 56
bootstrap 43, 52, 54, 58, 177

censored data 27, 67, 129
completely at random 29
components of cost-effectiveness 165
confidence box 94
confidence intervals 2, 44, 46
consistent estimation 30, 32, 34, 42, 44, 119
cost histories 28, 29, 31, 131
cost minimization 69, 102
cost prediction models 19
cost-effectiveness acceptability curve 9, 43, 52, 67, 72, 76, 81, 86, 90, 162, 163
cost-effectiveness plane 5, 6, 9, 43, 45, 60, 94, 95
covariate adjustment 10, 117, 147
covariates 14, 18, 24, 117, 163

decision-analytic modelling 165
direct (Lin) method 9, 28, 29
duration of interest 3, 11

EM algorithm 156
empirical Bayes 156
EQ5D 170
excess zeros 12, 18
expected value of perfect information 110–112
expected value of sample information 109, 112, 113

Fieller's theorem 46, 51, 54

generalized least squares 120, 125
generalized linear models 17

health related quality of life 176
heterogeneity 143, 152
hierarchical modeling 154, 162

incremental cost-effectiveness ratio 5, 44
incremental net benefit 2, 7, 8, 43, 49, 57, 67
informative censoring 27, 28
informative priors 89, 108
inverse probability weighting 9, 28, 31, 37, 39, 41, 78, 83, 129

kinked thresholds 69

life-table methods 9, 28, 39, 78
lost to follow-up 11, 27

mean quality-adjusted survival time 4,
    12, 22, 39, 133
mean survival time 3, 12, 21, 36, 134
multicenter trials 145
multinational trials 145, 180
multinomial regression 172

Newton-Raphson algorithm 156
non-censored data 11, 67, 117
non-linear 16
number-needed-to-treat 3

opportunity loss 109–115
optimal sample size 112–115
ordinary least squares 17, 119, 125

parameters of interest 2, 67
posterior distributions 52, 53, 57, 59,
    62, 65, 89, 90, 106, 107, 111, 112,
    156
power 93
price weights 1, 4, 158
probability of surviving 21, 30, 93
product limit method 30, 33, 132

QALY 6, 12, 20, 22, 23, 39, 88
quality of life 4, 22, 72, 132, 148, 166,
    176

random effects models 146, 154, 160

regression models 10, 14, 17, 19, 20,
    24, 117, 151, 170
restricted maximum likelihood 156
restricted mean survival time 3

sample size 93
seemingly unrelated regression 117,
    119, 125
sensitivity analysis 1, 8, 50, 60
shrinkage estimator 156, 157, 160
skewness 12, 13, 14, 15, 48, 147,
    163
smearing 16
statistical modeling 165
structural assumptions 179
sub-group analysis 86, 117, 167, 181
survival function 3, 4, 30, 31, 34, 36,
    181

Taylor series 16, 45, 48
threshold WTP 6, 7, 43, 44, 52, 60, 76,
    94, 100
trade-off quadrants 5, 48
transformations 12, 15, 152
two-part models 18
type I error 93, 104, 108, 114, 146
type II error 86, 104, 109

uninformative priors 52, 57, 59, 89
utility 4, 12, 20, 110, 169, 176

value of information 93, 116

willingness-to-accept 60
willingness-to-pay 7, 52, 67, 124

Statistics in Practice

Human and Biological Sciences

Berger – Selection Bias and Covariate Imbalances in Randomized Clinical Trials
Brown and Prescott – Applied Mixed Models in Medicine
Chevret(Ed) – Statistical Methods for Dose-Finding Experiments
Ellenberg, Fleming and DeMets – Data Monitoring Committees in Clinical Trials:
A Practical Perspective
Lawson, Browne and Vidal Rodeiro – Disease Mapping with WinBUGS and
MLwiN
Lui – Statistical Estimation of Epidemiological Risk
*Marubini and Valsecchi – Analysing Survival Data from Clinical Trials and
Observation Studies
Parmigiani – Modeling in Medical Decision Making: A Bayesian Approach
Senn – Cross-over Trials in Clinical Research, Second Edition
Senn – Statistical Issues in Drug Development
Spiegelhalter, Abrams and Myles – Bayesian Approaches to Clinical Trials and
Health-Care Evaluation
Whitehead – Design and Analysis of Sequential Clinical Trials, Revised Second
Edition
Whitehead – Meta-Analysis of Controlled Clinical Trials
Willan and Briggs – Statistical Analysis of Cost-effectiveness Data

Earth and Environmental Sciences

Buck, Cavanagh and Litton – Bayesian Approach to Interpreting Archaeological
Data
Glasbey and Horgan – Image Analysis in the Biological Sciences
Helsel – Nondetects and Data Analysis: Statistics for Censored Environnmental
Data
McBride – Using Statistical Methods for Water Quality Management
Webster and Oliver – Geostatistics for Environmental Scientists

Industry, Commerce and Finance

Aitken and Taroni – Statistics and the Evaluation of Evidence for Forensic
Scientists, Second Edition
Balding – Weight-of-evidence for Forensic DNA Profiles
Lehtonen and Pahkinen – Practical Methods for Design and Analysis of Complex
Surveys, Second Edition
Ohser and Mücklich – Statistical Analysis of Microstructures in Materials Science
Taroni, Aitken, Garbolino and Biedermann – Bayesian Networks and
Probabilistic Inference in Forensic Science

*Now available in paperback

Printed and bound by CPI Group (UK) Ltd, Croydon, CR0 4YY

27/10/2024

14580286-0001

Printed and bound by CPI Group (UK) Ltd, Croydon, CR0 4YY

27/10/2024

14580286-0001